# THE EXPERIENTIAL TAXONOMY
## A NEW APPROACH TO TEACHING AND LEARNING

# EDUCATIONAL PSYCHOLOGY

Allen J. Edwards, Series Editor
*Department of Psychology*
*Southwest Missouri State University*
*Springfield, Missouri*

*In preparation*

Gary D. Phye and Daniel J. Reschly (eds.). School Psychology:
Perspectives and Issues

*Published*

Norman Steinaker and M. Robert Bell. The Experiential Taxonomy: A New Approach to Teaching and Learning

J. P. Das, John R. Kirby, and Ronald F. Jarman. Simultaneous and
Successive Cognitive Processes

Herbert J. Klausmeier and Patricia S. Allen. Cognitive Development of Children and Youth: A Longitudinal Study

Victor M. Agruso, Jr. Learning in the Later Years: Principles of
Educational Gerontology

Thomas R. Kratochwill (ed.). Single Subject Research: Strategies
for Evaluating Change

Kay Pomerance Torshen. The Mastery Approach to Competency-Based Education

Harvey Lesser. Television and the Preschool Child: A Psychological Theory of Instruction and Curriculum Development

Donald J. Treffinger, J. Kent Davis, and Richard E. Ripple (eds.).
Handbook on Teaching Educational Psychology

Harry L. Hom, Jr. and Paul A. Robinson (eds.). Psychological
Processes in Early Education

J. Nina Lieberman. Playfulness: Its Relationship to Imagination
and Creativity

Samuel Ball (ed.). Motivation in Education

Erness Bright Brody and Nathan Brody. Intelligence: Nature,
Determinants, and Consequences

*The list of titles in this series continues on the last page of this volume*

# THE EXPERIENTIAL TAXONOMY

## A NEW APPROACH TO TEACHING AND LEARNING

*Norman W. Steinaker*

*M. Robert Bell*

ONTARIO–MONTCLAIR SCHOOL DISTRICT
ONTARIO, CALIFORNIA

ACADEMIC PRESS  New York  San Francisco  London  1979
*A Subsidiary of Harcourt Brace Jovanovich, Publishers*

COPYRIGHT © 1979, BY ACADEMIC PRESS, INC.
ALL RIGHTS RESERVED.
NO PART OF THIS PUBLICATION MAY BE REPRODUCED OR
TRANSMITTED IN ANY FORM OR BY ANY MEANS, ELECTRONIC
OR MECHANICAL, INCLUDING PHOTOCOPY, RECORDING, OR ANY
INFORMATION STORAGE AND RETRIEVAL SYSTEM, WITHOUT
PERMISSION IN WRITING FROM THE PUBLISHER.


ACADEMIC PRESS, INC.
111 Fifth Avenue, New York, New York 10003


*United Kingdom Edition published by*
ACADEMIC PRESS, INC. (LONDON) LTD.
24/28 Oval Road, London NW1 7DX


Library of Congress Cataloging in Publication Data

Steinaker, Norman.
    The experiential taxonomy.

    (Educational psychology)
    Bibliography: p.
    1. Learning, Psychology of.  2.  Experience.
I.  Bell, M. Robert.  Date      joint author.
II.  Title.
LB1051.S71255      370.15'2      79–1141
ISBN  0–12–665550–2


PRINTED IN THE UNITED STATES OF AMERICA

79 80 81 82     9 8 7 6 5 4 3 2 1

*TO* BLANCHE AND MARGARET

*and*

DR. HERSCHELL D. RICE
*whose influence on this work
is greater than he knows*

# Contents

*vii*

# Preface

The experiential taxonomy, developed in 1975 and tested and researched in the ensuing years, is an effective tool for teacher training, for teacher self-evaluation, and for curriculum development. When it is keyed in a curriculum to a series of taxonomically sequenced teaching strategies and learning experiences, it can augment learner achievement. Using the experiential taxonomy, one can plan an experience with specific objectives, with a series of taxonomically ordered activities keyed to identified teaching strategies, and with correlated elements of creativity, critical thinking, and problem solving. Furthermore, one can taxonomically order assessment strategies and key the whole package as a curriculum unit or learning experience that, so organized, can be an effective tool for increasing learner achievement.

A teacher using the curriculum package can audio- or videotape a selected lesson and, with a brief in-service of no more than 2 days, can code the lesson and analyze its effectiveness in helping students internalize the planned objectives. The teacher who knows the taxonomy and is coding hierarchically sequenced teaching strategies can understand why a lesson succeeded or why it did not. The teacher can begin to sense personal strengths and limitations as a teacher and can plan and participate in professional development accordingly. In short, the experiential taxonomy can serve as an in-service basis for teacher growth as a professional. Its efficacy for teacher training, for in-service, for self-evaluation, and for curriculum planning, implementation, and evaluation are immeasurable.

What is the experiential taxonomy, and why can we make these claims? In essence, it is a deceptively simple yet pedagogically and psychologically sound instrument. In the following pages, we will establish the need for a taxonomy of experience, trace its development, and show its usefulness as a model for curriculum development through a step-by-step pattern for experience design. We will, furthermore, show how it relates to learning principles and to creativity, critical thinking, and problem solving.

With this theoretical base, we will then cite developed examples of experiential-taxonomy-based curricula, showing how they were developed and how they can be implemented. The next chapter defines the teaching strategies in taxonomic order and explains the use of the taxonomy as a curriculum development tool. Curriculum implementation is then discussed along with evaluation and assessment procedures.

The taxonomy is, however, useful for more than curriculum development, implementation, and evaluation. An important function of this book is to show how the taxonomy is used in education in a variety of situations and settings. One chapter contains a taxonomic role model for teachers. This is followed by a chart of the taxonomically sequenced teaching strategies, with brief descriptors. Then follows a modus operandi for taping and coding a learning experience. This includes a step-by-step service program for teachers. Finally, this chapter contains some generalizations about the nature and the sequence of appropriate lessons, along with analyses of the interpretation of coded lessons. These generalizations are based on research and on extensive coding of classroom experiences. With the data in this book, a teacher can plan, implement, and evaluate learning experiences and, thus, can grow professionally.

It should be noted that this is a work unique in the field in that it was theoretically based. It is, therefore, a statement of educational and learning theory that, although it relates to other works in the field, is an essential work developed within a particular context. The research conducted was within that context and is referred to here. The book stands on its own merits as a unique expression both of theory and of research and practice. Readers will, therefore, not expect extensive references and bibliographical entries. In this sense, we make both a statement and a challenge. We invite our colleagues to research and to test this statement within other contexts so that the experiential taxonomy as stated can indeed become a useful instrument for measuring and understanding human experience.

We posit that the experiential taxonomy also has great worth in fields other than that of education. In the final chapter we explore many ways in which the taxonomy can be used. We suggest that any subjective communication or message in any form use the experiential taxonomy in identifying our objectives and in analyzing and interpreting our format and sequence of activities. It is furthermore suggested that the experiential taxonomy can be a major organizing factor in any campaign or effort aimed at changing behaviors or expected outcomes. In short, the experiential taxonomy can become a useful instrument not only in education but also in many other areas of human activities.

# Acknowledgments

This book is the culmination of many long months of thought, study, preparation, research, and evaluation. We have been most fortunate to have had the support, cooperation, and encouragement of numerous colleagues and friends, without whom this work could not have come to fruition. We must mention specifically Dr. Marilyn Harrison, who, as research associate, was of tremendous help in the research associated with the taxonomy. Her skills and insights helped bring many things together in this work. We must, likewise, credit Dr. Morris Krear of the California State Department of Education for recognizing early the viability of the taxonomy and for challenging and encouraging us to develop its new dimensions.

From Ontario–Montclair School District, Dr. Jack Jones, Superintendent, Dr. Jack Hassinger, Special Projects Director, and Dr. Lucile Robinson, General Consultant, offered great encouragement and support as the taxonomy developed. Their skills and distinctive points of view aided us in developing and editing this book. We must also express our thanks to the teachers who participated in the research projects and who helped develop the curriculum format. Our thanks to Aubrey Irwin, Bill Morgan, Bruce Perry, Steve Tucker, Carol Metzger, Lois Garrison, and Priscilla Ivory.

For reviewing the manuscript, for acing as special consultants, and for offering suggestions for its improvement, we want to express our appreciation to Dr. Everett Hilliard of West Covina Unified School District, to Dr. Margaret Lenz and Dr. Bob West of California State College, San Bernardino, and to Dr. Forrest Harrison of San Francisco State University.

Finally, we must express our appreciation to our staff members, whose typing, correcting of punctuation and spelling, and extra effort brought the manuscript into its final form. We thank Judy Cuthbertson, Linda Vaught, Louise Scritchfield, and Pat Nelson for their exceptional work, their patience, and their loyal support.

# THE EXPERIENTIAL TAXONOMY
A NEW APPROACH TO TEACHING AND LEARNING

# 1 | Introduction

**E**ducators, psychologists, school administrators, and consultants have, in the past years, been working with various taxonomies for planning, implementing, and evaluating classroom activities, student learning, and human experiences. Several taxonomies, including the *cognitive taxonomy* of Bloom, Englehart, Furst, Hill, and Krathwohl (1964), the *affective taxonomy* of Krathwohl, Bloom, and Masia (1968), and the *psychomotor taxonomies* of Harrow (1972) and Simpson (1966) have been published. These have proven to be useful tools for curriculum planners and for teachers in their work with students. Each of the taxonomies has contributed to more effective planning, implementation, and evaluation. Each has also had implications for teacher training and teacher evaluation. Yet each of these taxonomies addresses only one aspect of human experience. Although one can look at human experience through the cognitive taxonomy, through the affective taxonomy, or through a psychomotor taxonomy, none of these addresses the whole of human experience. Each deals with only a particular aspect of the total human interaction with an experience. A gestalt taxonomy is needed to provide a framework for understanding, planning, and evaluating the meaning of a total experience. Rather than using three or more taxonomies as models for planning, implementing, and evaluating, the educator in the field needs a taxonomy that is unified, complete, functional and can be easily used. In this book, we have responded to this need by proposing a taxonomy of experience that can be used for such purposes without dividing human experience into units or categories, partialities, or particular domains.

This, our view, is essential because, by definition, in *Webster's* (Webster's New World Dictionary of the American Language, second college edition), experience is an actual "living through an event or events." The "living through" of an experience involves the total personality. We suggest that an experience cannot be understood by fragmentation or isolation; it has identity, continuity, and a broad base involving all human senses and activities. For example, upon reflection one can cite stimuli that evoked the possibility of an experience. One can trace activities within the scope of an experience that sequentially brought about participation in the experience and, finally, dissemination of that experience, whether it was positive or negative. It is our contention that individuals, when they think of events from their own past, think of the totality of an experience, of the sequence of related activities within the experience, and of their involvement in those activities. Individuals think of experience as an integrated whole involving mind, physical being, and the sum of their previous experience. An experience involves all the extant classifications identified in the various cognitive, affective, and psychomotor taxonomies, but, when we remember an experience, we think of it as a whole entity and not as cognitive, attitudinal, or physical responses. The experience was all of these and more. It is within this concept of a total human experience that educators, psychologists, counselors, administrators—indeed, all people—must plan if they are to work more effectively with total personal involvement in a planned sequence of activities and are to be able to evaluate then the impact of individual interaction in that sequence of activities.

Our taxonomy was developing for a long time. For a number of years, after working with the existing taxonomies and utilizing their systems for planning and evaluating programs, we felt the need for a more broadly based taxonomy that could speak to the totality of an experience. The problem that finally brought the taxonomy into being was the challenge of a new program. As the needs of this new program were considered, we began to think in terms of what human experience really was and of how it could be categorized and brought into our planning. After a time of serious consideration and after periods of experimenting with various categories, the taxonomy took shape and has remained essentially the same since it was first proposed. The taxonomy has five basic categories and a number of subcategories. Whereas the total taxonomy will be examined in more detail later, the five basic categories are

       1.0  Exposure
       2.0  Participation
       3.0  Identification
       4.0  Internalization
       5.0  Dissemination

Through these categories a natural and logical progression is possible, leading to the planned outcome, namely, learning. Although they can stand alone and have some individual integrity, they are intrinsically linked together and, as previously indicated, an individual can move through an experience from exposure to dissemination, whether the experience is positive or negative. In the philosophical sense, we are axiologically neutral; in an educational sense, however, we feel that a planned learning experience should have definite outcomes and that these outcomes can be planned, implemented, and evaluated using this taxonomy. We furthermore believe that teachers using the experiential taxonomy can develop professionally through in-service sequences and through self-evaluation of the teaching–learning act using the experiential taxonomy.

In our present school district, the taxonomy has been used in a number of ways. Initially, it was used in the district career education program in which, within the context of a sequential program from kindergarten through grade 8, curriculum was developed based on the experiential taxonomy and was then successfully tested through the five categories through internalized and disseminated learning. It has proven to be successful not only in terms of positive reaction by the students and teacher but also in measured evaluations. A sequence of teacher in-service, curriculum units, and student activities related to the experiential taxonomy has greatly augmented the program and has developed teacher and staff use of and effectiveness with taxonomy-based materials. There is now operational, within the district, a whole series of sequential units based on the taxonomy in career education.

Another design based on the experiential taxonomy was a research project in the education of the handicapped, funded by the California State Department of Education. Here the major thrust was curriculum development. We worked with a team of educators including research associate Dr. Marilyn Harrison and the teachers involved in developing a series of curriculum units in reading, language, math, written communication, and social development for educationally handicapped (EH) and educable mentally retarded (EMR) youngsters. We also developed evaluation materials. The curriculum materials achieved remarkable success in the actual classroom. This research project showed the efficacy of the taxonomy-based curriculum, which proved to be as effective as or more effective than any other sequenced and structured curriculum. The Steinaker and Harrison study concluded that

> A carefully structured instructional program does contribute to student development. The Experiential Taxonomy and curriculum represent one such instructional program which enables a student to attain significant achievement. Such sequential instructional programs overcome limitations in student learning modes, achievement levels, educational and home background characteristics. The importance of a structured curriculum and instructional program does emerge as one of the most important factors, if not the most important factors, in determining whether a student does achieve as expected   [1976, pp. 75–76].

A research project associated with career education tested and found highly successful the curriculum development model as well as the teacher training, the in-service, and the evaluation models. It was during the career education research project that the associated hierarchy of teaching strategies, the finalization of the curriculum model, and the coding system for teacher self-evaluation were completed. The teaching strategies (Chapter 8) were used as the basis for analyses of the teaching–learning act, using the Experiential Taxonomy Teaching Strategies Coding System (Chapter 9). Teacher in-service on the coding system was instituted, and the study that validated the coding system through studies of interrater reliability was undertaken.

To determine inter-rater reliability, correlation analysis was used to determine the degree of association existing between the two raters for each taxonomic level. A correlation coefficient of .85 ($p < .01$ with 9 $df$) was obtained for level one and .92 ($p < .01$ with 99 $df$) for level two, indicating significant inter-rater reliability.

Validity depends on whether what was coded did in fact exist and whether these elements of the original situation were recreated in their proper perspective during the decoding process. It was judged that the strategies do accurately reflect the complexities of the classroom. Further, the Experiential Taxonomy Teaching Strategies Coding System is an objective instrument allowing a correct recreation of the classroom [Steinaker & Harrison, 1977, pp. 49–50].

The study further supported "the case for using curriculum and teaching strategies keyed to the Experiential Taxonomy [Steinaker & Harrison, p. 56]."

The taxonomy is presently being used in developing materials in other areas as well. Multicultural curriculum units and teaching sequences based on the experiential taxonomy have been developed for use in Grades K–8. A series of simulations are being developed for the junior high school level. Another interesting in-service activity based on the experiential taxonomy was the initial in-service for the district's management team in preparation for collective bargaining that went into effect in mid-1976. All these, as well as other projects and activities, have sustained our belief that the taxonomy has real usefulness for anyone planning experiences in any academic field and for teachers serious about self-evaluation and about professional development. The practicality of the experiential taxonomy is demonstrated throughout this book, but in particular the reader should note those sections that deal with curriculum development and with evaluation (Chapters 4 and 9) and those dealing with teacher change and with professional development (Chapters 10 and 11).

Shortly after the development of the taxonomy, an article was submitted to *Educational Technology*. The article, "A Proposed Taxonomy of Educational Objectives: The Experiential Domain," was published in January, 1975, and shortly thereafter inquiries and letters from colleagues around the world were received. The great variety of responses to the article was surprising. Whereas communications from our colleagues in education were expected, those from medical professionals, counselors, psychologists, and psychiatrists, as well as those from sociologists, theologians, and penologists, were not. Each succeeding article has elicited equally wide response. The comments, reactions, and requests for further information indicated anew the need for this kind of taxonomy and gave us the incentive to develop the taxonomy further, extending and expanding its usefulness in teaching and producing this book.

No work of this nature is really successful, however, unless the model functions effectively in the real world of teaching and learning. Theory means little unless it can be translated into practice. In this book, we will present a series of practical applications of the theoretical models, along with the effects of those applications. In conjunction with the functional curriculum examples presented in this book, we will discuss the use of the experiential taxonomy in staff development and in-service training not only for teachers but also for parent groups, for citizens, and for others.

We feel very strongly that this taxonomy can have an impact on planning and evaluating human experiences in all segments of life. If readers are responsible for planning, implementing, or evaluating activities, the experiential taxonomy can be a real asset as they begin to innovate in their own area of responsibility.

# 2 | Building an Experiential Approach to Learning

In any reflection upon experience itself, a corollary question comes to mind. Is there a hierarchy of experience, a sequence of steps or of definable categories through which an individual proceeds or progresses from the initial exposure to an experience through the dissemination of that experience? As one ponders past personal experiences, the answer certainly becomes affirmative. There is, indeed, a related continuum of perceived and actual experience, beginning with the sensing of the possibility of a deeper, more complex experience and moving on sequentially and logically toward dedicated dissemination of that experience.

These perceived sequences of experience have never before been taxonomically organized. When existing taxonomies of educational and psychological objectives and their relationships to classroom activities, teaching strategies, counseling techniques, and degrees of

7

human experience are reviewed and used, one notes that each of those taxonomies focuses on only one dimension of the total human experience. Unfortunately, there has been, up to now, no taxonomy of educational objectives utilizing the experiential domain available to teachers, to counselors, to curriculum planners, and to evaluators. There is a definite weakness in planning classroom and human experiences, assessing changing behavior, and developing activities leading to taxonomically defined objectives without a complete encompassing of human activity in an experience in the planning. This weakness is further compounded when one tries to evaluate systematically the process involved in a learning experience or to provide for teacher growth and self-evaluation. There is a functional need for an experiential taxonomy to fill these gaps in the total planning and evaluation process. This is true not only in the classroom but also in the counseling session, in administration and business activities, and in personal as well as family activities.

Experience itself needs a working definition at this point. As mentioned on page 2, *Webster's* defines it as "an actual living through of an event or events." This statement can serve as a beginning operational definition of experience. We know from our personal past activities that human experiences are inextricably linked together and that people bring to each experience the sum total of their previous activities. Yet, when we as individuals begin something, we know also that there is going to be some continuity and identity to the event or events of an experience. We can reflect on the stimuli that indicate the possibility of a deepening and ongoing experience and may then trace our progress through some participation in the event to an identification with it. Various aspects of the event or events that have had particular impact then becomes internalized in our personality and behavior. Finally, we relate this experience to our previous experiences and to other individuals. This step is the dissemination of an experience.

Experience can be, of course, either positive or negative. Every individual reacts to or responds to an experience either positively or negatively. Whatever the reaction, one can still move through all steps of an experience from exposure to dissemination. Everyone knows of or has been involved in a negative experience and its dissemination. When one really considers all dimensions of an experience, the result of the dissemination is immaterial. We are, in effect, axiologically neutral when experience is discussed.

Experience also either involves the individual physically or is vicarious. The problem of differentiating these kinds of experiences may arise as one begins planning, implementing, and evaluating experiences. In our view, there is no taxonomic difference between an experience in which one is physically involved or one in which there is vicarious involvement. Human beings are total creatures, and our mental and physical selves cannot be differentiated or separated from each other. One needs only to turn to the creative process or to reading and television to see how deeply felt vicarious experiences can be. In a vicarious experience, an individual can move through all the categories of an experience, from exposure to dissemination. In this taxonomy, we take the ontological position that all experience is real to the one who actually or vicariously lives "through an event or events." In terms of the taxonomy, we do not accept any ontological position that tends to define and to limit reality to those kinds of activities involving only physical experiences. In essence, we see no taxonomic difference between a vicarious experience and an experience in which one is physically involved. In counseling situations or in psychology, however, the counselor may wish to use the taxonomy for planning situations and interactions in which a counselee needs to differentiate between these two kinds of experiences. This, too, is taxonomically possible.

For purposes of the taxonomy, however, one does not need to be concerned with a definition of an experience as real or vicarious or even as fantasy. Rather, teachers, counselors, curriculum planners, and/or evaluators can, with a brief motivation (exposure) and orientation (participation), begin to work with individuals at whatever category of the experiential taxonomy the individual is reflecting at a particular time. They can, furthermore, plan additional activities that will enhance the experience or will change its direction and outcome. For our purposes, then, experience can be defined as a hierarchy of stimuli, interaction, activity, and response within a scope of sequentially related events beginning with exposure and culminating in dissemination. Experience can be either positive or negative, and the value orientation of an experience can extend through the dissemination. Not every stimulus in our lives is followed up, nor is everything in which we participate. One can reject or terminate an experience at any level of the taxonomy, from exposure through identification and internalization. Dissemination of an experience, particularly if it is negative or distasteful or involves anxiety, fear, or guilt, may not occur at all. One must also remember that dissemination is a learner-chosen activity. We will later discuss teacher role models for each level of the taxonomy. For counselors and psychologists, each step in the taxonomy is of equal importance, but, for lasting learning, the identification level and those following are critical. It is at these levels that a counselee, for example, can bring out the past and can begin to examine it.

With these thoughts and with the previous definition of experience, we propose the following taxonomy of educational objectives: the experiential domain.

### Categories of the Experiential Taxonomy

1.0 EXPOSURE: Consciousness of an experience. This involves two levels of exposure and a readiness for further experience.

    1.1 SENSORY: Through various sensory stimuli one is exposed to the possibility of an experience.

    1.2 RESPONSE: Peripheral mental reaction to sensory stimuli. At this point, one rejects or accepts further interaction with the experience.

    1.3 READINESS: At this level one has accepted the experience and anticipates participation in it.

2.0 PARTICIPATION: The decision to become physically a part of an experience. There are two levels of interaction within this category.

    2.1 REPRESENTATION: Reproducing, mentally and/or physically, an existing mental image of the experience, that is, through visualizing, role playing, or dramatic play. This can be done in two ways:

    2.1.1 *Covertly:* as a private, personal "walk-through" rehearsal.

    2.1.2 *Overtly:* in a small or large group or interaction, that is, in the classroom or playground.

    2.2 MODIFICATION: With the input of past personal activities, the experience develops and grows. As there is a personal input in the participation, one moves from role player to active participant.

3.0 IDENTIFICATION: The coming together of the learner and the idea (objective) in an emotional and intellectual context for the achievement of the objective.

3.1 REINFORCEMENT: As the experience is modified and repeated, it is reinforced through a decision to identify with the experience.

3.2 EMOTIONAL: The participant becomes emotionally identified with the experience. It becomes "my experience."

3.3 PERSONAL: The participant moves from an emotional identification to an intellectual commitment that involves a rational decision to identify.

3.4 SHARING: Once the process of identification is accomplished, the participant begins to share the experience with others, as an important factor in his life. This kind of positive sharing continues into and through Category 4.0 (internalization).

4.0 INTERNALIZATION: The participant moves from identification to internalization when the experience begins to affect the life-style of the participant. There are two levels in this category.

4.1 EXPANSION: The experience enlarges into many aspects of the participant's life, changing attitudes and activities. When these changes become more than temporary, the participant moves to the next category.

4.2 INTRINSIC: The experience characterizes the participant's life-style more consistently than during the expansion level.

5.0 DISSEMINATION: The experience moves beyond internalization to the dissemination of the experience. It goes beyond the positive sharing that began at Level 3.0 and involves two levels of activity.

5.1 INFORMATIONAL: The participant informs others about the experience and seeks to stimulate others to have an equivalent experience through descriptive and personalized sharing.

5.2 HOMILETIC: The participant sees the experience as imperative for others.

It should be noted that the categories in this taxonomy are stated in positive terms, even though, as noted earlier, an experience can elicit either a positive or a negative reaction. For purposes of educational planning, for developing sequential counseling experience, or for planning in any area of human experience, those involved in planning, implementing, or evaluating the experiences usually see their goals in positive terms. A planner seldom, if ever, plans experience in which those involved are expected to have negative reactions. It is for this reason that the categories are stated positively.

# RELATIONSHIP TO OTHER TAXONOMIES

The experiential taxonomy as a gestalt approach to human experience relates closely to existing taxonomies of educational objectives. As one reviews the sequence of experience as noted in the taxonomy, one sees an immediate relationship to the affective taxonomy of Krathwohl *et al.* Exposure in the experiential taxonomy at the sensory (1.1) level relates to the awareness (1.1) and willingness-to-receive (1.2) categories of the affective domain, and the response (1.2) and readiness (1.3) categories of exposure encompass the controlled or selected attention (1.3) and acquiescence in responding (2.1) categories of the affective taxonomy. This relationship between the two taxonomies continues as one compares the various categories. The experiential taxonomy is perhaps most closely related to the affective taxonomy, particularly at the identification (3.0) and internalization (4.0) categories, where it parallels the organization (4.0) and characterization (5.0) categories with a value or value complex of the affective domain.

It should be noted here that the experiential taxonomy, since it is a classification of a total experience in all its variant facets, augments the affective taxonomy as well as other extant taxonomies. It particularly augments the affective taxonomy by defining the motivational and interactive aspects of an experience in the exposure (1.0) and participation (2.0) categories, which bring the individual to an identity with an experience. It also provides a follow-through in a sequence of ever-deepening activities that define experience beyond Category 5.0 of the affective taxonomy (characterization by a value or value complex). The dissemination category (5.0) of the experiential taxonomy, by including the choice to relate the event or events to others, adds another dimension to human experience. This level of human activity is neither implied nor noted in the categories of the affective taxonomy. The experiential taxonomy, therefore, provides a much broader perspective, a more logical organizational framework, and a more complete sequence of human activity than does the affective taxonomy.

The experiential taxonomy likewise supplements and strengthens the cognitive taxonomy of Bloom *et al.* by making explicit the classification of sequences of activities that will bring people to the highest levels of that taxonomy. Each experience planned and based on the categories of the experiential taxonomy relates to the cognitive taxonomy. Knowledge (1.0) and comprehension (2.0) in the cognitive taxonomy require exposure (1.0 in the experiential taxonomy) through various motivational strategies that are strong at the exposure level of the experiential taxonomy and require also a follow-up in the participation level (2.0) of the experiential taxonomy. It is at this level of the experiential taxonomy that learners begin to understand the parameters of the experience. Application (3.0) in the cognitive taxonomy involves some aspects of participation (2.0 in the experiential taxonomy) and Sublevels 3.1 and 3.2 of identification (3.0), at which time the learner uses and begins to apply data. The higher levels of the cognitive taxonomy, analysis (4.0), synthesis (5.0), and evaluation (6.0), would closely parallel the hypothesizing and testing aspects of identification (3.0) and the internalization category (4.0) of the experiential taxonomy. Dissemination (5.0) of the experiential taxonomy, again, carries human activity beyond the cognitive taxonomy's domain. Whereas the cognitive taxonomy remains a useful tool for teacher planning of the teaching–learning act, it too is limited in that it does not relate to the whole of human activity or experience. It must be augmented by other taxonomies. The experiential taxonomy does not require such augmentation; rather, it sees the sequence of human experience as a whole entity and defines it through dissemination (5.0), which the cognitive taxonomy only infers but does not make explicit.

Although her construct of the psychomotor activities differs in sequence from that of the experiential taxonomy in some categories, the psychomotor taxonomy of Elizabeth Simpson (1966) corresponds to the experiential taxonomy. Her category of perception (1.0), with its subheading, sensory stimulation (1.1), is very similar in definition to the exposure category (1.0) of the experiential taxonomy, particularly the subcategory sensory exposure (1.1). Category 2.0 (set) in Simpson still relates to exposure, though her Elements 2.1 and 2.2 parallel the elements of identification, specifically 3.2 and 3.3. Guided response (3.0) relates to some elements of participation (2.0) as well as to elements of identification (3.0). Her trial and error category (3.2) would relate to teaching strategies and learning experiences in Categories 3.3 and 3.4 of the experiential taxonomy. Mechanism (4.0) in Simpson is close to some elements of internalization, as in her Category 5.0, complex overt response. Again, she does not consider dissemination.

Thus, one can see that the experiential taxonomy is related to the other taxonomies in many respects; the experiential taxonomy, however, provides a more complete vehicle for planning, for sequencing, for implementing, particularly for evaluating human experiences in terms of the teaching–learning act, because it deals with experience as a total human activity rather than dividing that activity into its cognitive, affective, and psychomotor dimensions. The experiential taxonomy can, perhaps, become the most useful of all taxonomies for realistically planning, implementing, and evaluating educational objectives and the related teaching–learning act, because the experiential taxonomy provides a synthesis of the cognitive, affective, and psychomotor elements of the learning process. This synthesis encompasses all aspects of the other taxonomies and brings human experience together into a manageable frame of reference for functional use at both the theoretical and the practical levels of research, teaching, and learning. This taxonomy provides a way to deal with the totality of learning activities. It is an answer to the problem of defining the sequence of activities and feelings one follows from exposure to an experience through dissemination of the experience. It will, furthermore, make it easier for a teacher to plan objectives appropriate to the needs of an individual learner or groups of learners because the teacher can deal with total individual needs in a succinct and logical manner. It will also help teachers recognize and understand where various students are in terms of a particular experience and will thus enhance the teacher's ability to develop new strategies and activities designed to help each individual through additional levels of experience. Furthermore, it can be of vital importance in teacher self-evaluation and in professional development. When the teacher learns to use the taxonomy for self-evaluation, it becomes perhaps the most effective tool in professional growth. Other taxonomies do not deal with teacher self-development.

As has already been noted, one vital element of human activity that the experiential taxonomy encompasses within its construct and that the other taxonomies fail to consider is the dissemination of an experience. This is what one does with an experience. It is the living out of an internalized activity. Dissemination is the level at which one may choose to relate experience to others. Within the experiential taxonomy, one can plan for the implementation of dissemination and can evaluate the dissemination of experiences.

Using this taxonomy, a teacher can deal specifically with an experience or series of experiences, can identify the level of activity involved, and can evaluate personal effectiveness in teaching those activities. Furthermore, the teacher can structure experiences designed to move the learner from exposure to dissemination or to any intermediate level, depending on the needs of the learner. With the experiential taxonomy, the teacher will not only have filled a void untouched by the other taxonomies but will also have a format that sees cognition, evaluation, and psychomotor activity as a part of total learner experience. A learning experience can be planned using the experiential taxonomy as a basis for a format, and measurement instruments can be developed from it to assure and enhance a continuing experience and learning.

## THE EXPERIENTIAL TAXONOMY AND INSTRUCTIONAL THEORY

For a number of years, educators have discussed the relationship of instructional theory to educational practice (Gage, 1963, pp. 94–141). Numerous studies have indicated a tenuous, even remote, relationship between learning theory and educational practice. They have often seen instructional theory as a link between learning theory and actual teaching practices. Saettler, for example, suggests that instruction theories could fulfill a mediating role between basic learning theory and practice (Gage & Rohwer, 1969, pp. 381–418). Others, like Siegel (1967), find formal education to be "long on practice and short on theory." He also points out that it does make a difference how students are taught and that part of the problem in teaching is a relative lack of adequate theory. "A sound theoretical base can (1) suggest better educational practices than are now prevalent; (2) permit predictions about the likely effectiveness or ineffectiveness of contemplated innovations . . . ; and (3) guide future research efforts in systematic rather than fragmented directions [p. ix]."

Others, like Bruner (1964), feel that there must exist a theory of instruction as a "guide to pedagogy." Skinner (1968) argues for directly studying the teaching–learning process in order to develop principles for teaching. Glaser (1966, pp. 433–449) sees the need for the translation of theory into practice, and Ausubel (1968) has also pointed to the need for theories of instruction that take as their central concern meaningful school learning. With this emphasis on the need for sound and functional educational theory, a number of systems and theories have been developed, including Skinner's programmed instruction based on operant theory and techniques to improve instructional procedures. More recently, great interest has developed in teacher accountability, with a focus on outcomes or products and instructional research and theory.

Another recent emphasis is on classroom interaction analysis systems and instructional principles. More than 70 such systems have developed that examine what happens in the classroom and point out afresh the need for a practical instructional theory.

The curriculum reform movement has also been instrumental in stimulating educational theory. Educators are concerned with the theory, the design, and the improvement of curriculum and instruction. Though several groups have addressed this subject, one national organization especially concerned with curriculum reform is the Association for Supervision and Curriculum Development (ASCD). In the late 1950s and in the 1960s, a great interest in a systematic basis for curriculum developed. The ASCD, in response to this interest and recognizing the need for systematic curriculum development, established the Commission on Instructional Theory. The Commission, chaired by Ira J. Gordon, established (Gordon, 1968) the following criteria for instructional theories:

1. A statement of an instructional theory should include a set of postulates and definition of terms involved in these postulates.
2. The statement of an instructional theory or subtheory should make explicit the boundaries of its concern and the limitations under which it is proposed.
3. A theoretical construction must have internal consistency—a logical set of interrelationships.
4. An instructional theory should be congruent with empirical data.
5. An instructional theory must be capable of generating hypotheses.
6. An instructional theory must contain generalizations that go beyond the data.
7. An instructional theory must be verifiable.
8. An instructional theory must be stated in such a way that it is possible to collect data to disprove it.
9. An instructional theory not only must explain past events but also must be capable of predicting future events.
10. At the present time, instructional theories may be expected to represent qualitative synthesis [pp. 16–24].

The experiential taxonomy meets these criteria. It contains a set of postulates and defines the terms involved in the postulates (Criterion 1). Curriculum developed using the experiential taxonomy defines objectives behaviorally and can deal with both student and situational characteristics together, according to the defined objectives. The experiential taxonomy meets Criterion 2 by defining instructional processes taxonomically, in terms of teaching strategies, levels of critical thinking, problem solving, creativity, and learning conditions. Although some experiential taxonomy instructional strategies may be used more at one age level than another, age level is not considered a boundary or limitation.

Criteria 3 and 4 are met. There is an obvious internal consistency to the experiential taxonomy, and it is "congruent with empirical data." The experiential taxonomy likewise stands the test of Criteria 5 and 6. The taxonomy can indeed generate hypotheses that can be specifically defined and tested. The Experiential Taxonomy Teaching Strategies Coding System (ETTSCS) for evaluation of the teaching–learning act meets Criterion 6. The experiential taxonomy has been verified through research and can and should be tested in a variety of circumstances (Criterion 7). Data can also be collected to prove or disprove it (Criterion 8). Since the ETTSCS can be used to explain a prior teaching situation, it can be used to predict student performance and learning under prescribed conditions (Criterion 9). Finally, the experiential taxonomy does represent qualitative synthesis (Criterion 10).

Although the preceding statement is brief, it does indicate that the experiential taxonomy is an instructional theory with pragmatic application to the teaching–learning act. It meets and goes beyond the ASCD criteria. Research and teacher in-service have shown it to be understandable and practical for teachers in curriculum development, implementation, and evaluation, as well as in self-evaluation and professional growth.


## RELATIONSHIP OF THE EXPERIENTIAL TAXONOMY TO LEARNING THEORIES

Educators have long lamented the lack of substantive relationship between learning theory and actual classroom teaching practice. Although instructional theory provides a bridge, it is often not used by either practitioners or theorists. One strong advantage of the experiential taxonomy is that it is compatible with learning theory. Since it emphasizes the process of teaching as well as the product, it works with any learning theory. A psychologist, counselor, or teacher can, for example, identify a student as being at a particular level in Piaget's developmental stages (Helder & Piaget, 1958) and can then use the experiential taxonomy to plan learning activities compatible with the Piagetian developmental stages. The taxonomy works similarly with Bruner's theory, which advocates discovery learning and suggests that any child can learn any concept if it is presented in an appropriate way with appropriate methodology (Bruner, 1964). The experiential taxonomy can provide a format for developing and evaluating the methodology.

Ausubel (1968), in discussing the learning sequence, indicates the need for an "advance organizer" that links up with the learner's "cognitive structure" and provides the learner with the necessary "subsumers" that make subsequent learning "meaningful." In terms of the experiential taxonomy, Ausubel suggests that there must be Exposure (1.0) through the advance organizers and then Participation (2.0) that uses the learner's cognitive structure or past experience to establish Identification (3.0), when the learning begins to become meaningful. This meaningfulness continues through Internalization (4.0) until the learner is able to Disseminate (5.0). The relationship of Ausubelian theory to the experiential taxonomy is thus apparent.

Other learning theories apply similarly. With the experiential taxonomy, one has a versatile vehicle for theory, no matter what assumptions the theory makes about learning or human development. The experiential taxonomy serves, therefore, as an organizational format for building an experiential approach to learning. Learning theory in conjunction with the experiential taxonomy can become an effective element in testing and evaluating student achievement. Theory and practice can become one.

# 3 | Expanding the Taxonomy

The major purpose in developing the taxonomy was to provide one unified, complete, and easily followed organizational structure for educational planning, implementation, and evaluation. It was conceived as a functional vehicle for providing the complete classification of human activity from the moment the learner is exposed to the possibility of an experience to its highest level of completion. The taxonomy defines the sequence of activities and feelings that the learner follows from exposure to the dissemination of an experience. In addition, it enables the teacher to identify the level of experience attained and to plan necessary assistance for each individual's further achievement. It also expedites the implementation of the plan and the evaluation of the results.

The experiential taxonomy is not a complicated system. It is designed to be used peripherally as well as in depth. Peripherally, it can be used in teaching simple units, in developing learning packages, in selling a product, or in designing a speech. For maximum effectiveness in a variety of situations, whether these be educational, political, economic, or sociological, the planner is urged to use the taxonomy in depth by analyzing the possibilities of each subcategory. The sections that follow will present the taxonomy in several ways, including a simple *walk-through* of the taxonomical categories and subcategories, followed by a description of each classification, what it includes, what it implies for learners, for teachers, and for evaluators, and, finally, how it relates to the teaching–learning act.

We point out that we have not noted examples of objectives at each experiential level, as authors of other taxonomies have frequently done. We submit that within this taxonomy a learning experience objective is achieved at the identification subcategory sharing level (3.4). At this point in the sequence, the student begins to recognize that learning is taking place and that new sets of skills, behaviors, and attitudes are now available for further development. In essence, the teacher's evaluation prior to the achievement level is of the process type. For purposes of planning, the teacher needs to note that summative or product evaluation is not done until level 3.4 is reached (Bloom, 1971). One interesting aspect of the experiential taxonomy is that it measures both the initial learning at the identification stage and the more in-depth data study at subsequent levels. In fact, it is the only taxonomy that clearly continues to the dissemination level.


## AN ORIENTATION WALK-THROUGH

Before we discuss each level of the taxonomy, the reader may experience the continuum as we briefly illustrate the possible steps in learning a new song. Table 3.1 shows the taxonomic steps a learner could go through in this process. It is obvious that the learner of the song could have rejected the experience at any level and might not have completed the continuum. One should be equally aware, however, that, if the objectives were to teach the melody and words, a deliberate attempt by the teacher to present activities that ensured progress at least through Level 3.4 would be necessary to demonstrate that the objective had been achieved. Herein lies the effectiveness of the taxonomy. It provides a guide whereby the teacher can design not

**Table 3.1**

Taxonomic Sequence of a Learning Experience

| Taxonomic levels | Steps |
| --- | --- |
| 1.0 Exposure | |
|    1.1 Sensory | I hear the song. |
|    1.2 Response | I enjoy it. |
|    1.3 Readiness | I want to hear the song again. |
| 2.0 Participation | |
|    2.1 Representation | I attempt to reproduce the melody. |
|    2.2 Modification | I add my own emphasis or style. |
| 3.0 Identification | |
|    3.1 Reinforcement | I repeat the song and the style often. |
|    3.2 Emotional | It is now one of "my" songs. |
|    3.3 Personal | I prefer "my" version or as "I" feel I first heard |
|    3.4 Sharing |    it to others. |
| 4.0 Internalization | |
|    4.1 Expansion | Based on previous reactions, it is not just one of |
| |    my songs, it's "me" and I find opportunities to |
| |    use it. |
|    4.2 Intrinsic | The words and music now have a special mean- |
| 5.0 Dissemination |    ing to me that they may not have for others. |
|    5.1 Informational | I provide opportunities for others to use and |
|    5.2 Homiletic |    experience "my" version. |
| | I feel and act as though others must feel as "I" |
| |    do about "my" version. |

only teaching strategies and learner behaviors but also evaluation strategies based on the degree of experience attained.

It should be noted at this point that the taxonomy is designed to be used in all kinds of classes. Research has illustrated its effectiveness in special education (EMR and EH classes). Two projects in the Ontario–Montclair School system, Ontario, California—Project MEET (1976) and Project TORCH (1977)—have demonstrated equal effectiveness for slow, for average, and for accelerated students in heterogeneous classes. One should also note that there are critical roles and tasks that a teacher assumes at each taxonomic level. For example, there is a changing teaching role at each taxonomic level, and there are changing teaching strategies, learning conditions, and evaluation procedures appropriate to each taxonomic level. These will be touched on briefly in this chapter and explained more thoroughly in later sections of this book.

## 1.0 EXPOSURE

Every true teacher realizes that in most learning situations—in every new circumstance or activity that involves growth or change—there is an element of uneasiness for the student. This uneasiness, or fear of the unknown, is a necessary part of all experience because, if there is to be a new experience, the student will be following a course he or she cannot completely control and often cannot understand. Although some degree of anxiety may be unavoidable or even desirable, the art and science of teaching require that attention be devoted to this feeling at the start; otherwise, the student may never move beyond the exposure level.

Exposure is defined as the consciousness of an experience. The student moves from isolation to readiness and goes through all those behaviors characterized by sensation, emotion, volition, and thought. It requires one or more mental processes and involves "noticing" with at least a minimum degree of controlled thought. It differs slightly from awareness in that it implies, rather, attention and some discrimination.

In moving through the exposure category, we focus on two subareas that prepare the student for further progress in the continuum. As Piaget implies, learning is provoked by situations, and, thus, planning must consider both the sensory and the response subcategories if the student is to become ready for further involvement (Phillips, 1969). These two divisions operate so closely that it is impractical to discuss them separately.

Sensory involvement may best be described as the relationship between physical stimuli and the reaction they produce in the human observer. Although it is through various sensory stimuli that a person becomes exposed to the possibility of an experience, the concomitant mental reaction will determine whether one accepts or rejects further interaction with the experience. In the teaching–learning experience, in which the learner may or may not be able to retreat physically, the learner can fail to move to other levels if planning treats this stage lightly.

*IMPLICATIONS FOR TEACHERS, LEARNERS, AND EVALUATORS*

Any attempt to bring about growth or change in individuals imposes certain obligations on those charged with the task. The teacher's responsibilities at the exposure level are to motivate the student, to focus attention, and to keep the anxiety level within bounds in order to maintain student confidence in possible success. Here the teacher must impart to learners an understanding of the learning process, and enjoyment in learning and a sense of its importance (Rosenshine, 1976). The teacher selects activities that energize the learner, that awaken interest, need, or drive, and that provide a feeling of purpose. Depending on timing and learning styles, activities range from data presentation to demonstration and audiovisual approaches. Questions are of a naive type designed to provoke attention, to stimulate "wonder," and to promote purpose or problem identification, but they rarely demand an answer.

The teacher's role responsibility here is that of a motivator who works to eliminate competing agents by shifting modes of instruction and sensory channels; the teacher uses movement, gestures, and voice inflections and/or alters the physical aspects of the learning environment. In this sense the teacher is dominant in the taxonomic level process, whereas the learner is receiving the stimuli of motivation (1.1), responding to it (1.2), and preparing for further experience.

Most learners at this level will be involved in activities such as seeing, hearing, reacting, recognizing, smelling, tasting, and touching. Although some prerequisite work–study skills may be necessary, these are usually the basic "intake" skills required for attending.

Peripheral evaluation at this first level, as in the case of exposure to simple units of work, involves primarily observing and sensing the positive and/or negative reactions of the learners, gauging their initial understanding and willingness to proceed, and analyzing such results. It is less formal than rating scales or checklists; yet, it is directed in that it attends to such variables as timing, the nature and clarity of the presentation, the individual learners, and the conditions under which exposure was to occur. When exposure occurs over a longer period of time, as in more complex situations where the learner has more control, evaluators would be expected to use measures of attendance, the learner's self-evaluations, interviews, and similar measures for data collection.

### Examples of Exposure-Level Activities

- Using audio or visual materials to create a need, purpose, or desire
- Presenting examples or experiments to illustrate a new principle, concept, or skill

- Locating resources in one area to arouse interest
- Presenting facts or principles
- Initial viewing of scenes, objects, and roles
- Unstructured interviewing
- Viewing dramatic presentation
- Asking fundamental and often naive questions
- Changing the relationship between previously used words, pictures, or activities and their appearance, color, and loudness
- Causing an alerting reaction through any novel, difficult, or unusual occurrence

From this list, it can be seen that exposure is the invitation to an experience where, in the teaching–learning act, extrinsic forms of motivation are used to gain and focus attention, to reduce anxiety and competing behaviors, and to establish in the learner a willingness to participate further. It is, as stated earlier, a movement from isolation to that readiness that is necessary to establish initial rapport with the imminent learning experience.

## 2.0 PARTICIPATION

The learner moving into this level has already been introduced to the experience and is motivated to look further. It is here that a willingness to proceed increases in strength. The participation level serves as a necessary bridge to learning and is characterized by a conscious effort on the part of the learner to replicate in some way that to which the learner has been exposed.

Participation has been described as the level at which one decides, on the basis of data already received, to become physically a part of the experience. It has been divided into two subcategories to show how participation is reinforced by either external or internal forces. The first stage, representation, involves reproducing covertly or overtly the presently held image or picture. In the second stage, modification, the learner needs to clarify meanings and to relate the new to the previously learned data. Both of these stages need further explanation.

Representation is characterized by the feeling of discovery. It is reinforcing in that the learner more consciously reviews what has been presented and the tone is set for the rest of the experience. No matter how brief a stage, it is indispensable to the modification process. It is a mental or physical rehearsal that brings the new experience closer, helps to reduce uneasiness, and starts the learner on a new adventure.

Modification, though regulated somewhat by the learner's feelings, moves more toward cognitive verification. There is a further accumulation and rearrangement of data as the learner gives up some previous impressions. The learner sequences, establishes tentative hierarchies of material, and defines a beginning frame of reference during this stage. Thus, although there is a clear distinction between the discovery process of representation and the verification process of modification, both are essential for progress along the continuum.

## IMPLICATIONS FOR TEACHERS, LEARNERS, AND EVALUATORS

Although some necessary groundwork has been covered, the teacher's role as a catalyst is vital here for student progress to continue. The learner's acceptance of the teacher's guidance through the experience is now more crucial than before, since it begins to involve changes in behavior and in the perception of what is right for the individual. Much initial guidance and supportive feedback will reinforce a positive association with and additions to the experience data. The teacher should clearly indicate where learners are going, how to get there, and what it is going to look like when they arrive. As noted in a later chapter, in many instances this is where the learner begins to feel a need to know the rules within which to operate, what the obstacles are that must be overcome, and what rewards, benefits, advantages, abilities, and/or knowledge will be attained upon completion. Teachers should consciously incorporate student ideas into the lesson where possible yet should also maintain the prerogative of restructuring and of refocusing them on the lesson when listening to student responses. It is also here that the teacher must be able to promote further student thought about the data presented. Situations that emphasize recall or replay of data, sample use of material or information, and the connection of material by sequence or pattern should be presented to multiply the experiential base. Warm-up recall and translation questions become important at this level because they help learners bridge the gap between what they already know and what they need to know. At this stage appropriate questions from the teacher encourage students to profit from pulling together a common information base from which to operate and learn.

Learner behaviors during this phase involve mental and/or physical activity. Mental activities are imitative, such as visualizing, modeling, "walking through," or recalling a situation similar to that presented at the exposure level. In the physical activities the student begins to explore, manipulate, collect, and discuss the data available and to make certain inferences between them and what is already known. Finally, at this taxonomic level, the learner can be expected to make some figurative or symbolic response expressing these inferences, tentative though they may be. This expression could involve sequencing data, defining a frame of reference, or establishing a material hierarchy.

A specific evaluation problem begins to appear at this level, and it is one that must be weighed during the activity planning stage. The teacher more easily examines and judges the designed and implemented activities on the basis of external and overt behaviors. Many of the responses made by the learner, however, could be covert in nature, and therein lies the difficulty. Though the covert response is more efficient in terms of learning time, the evaluator still needs an overt indication of actual learning. The evaluator, therefore, at this level, is the questioner and must determine whether the learner's data acquisition calls for further advancement or for a change in the instructional plan to provide the necessary background. The teacher must ask questions that demonstrate understanding, ability to succeed, and, where appropriate, how the learner would "do it" if provided the opportunity. The teacher is, further, expected to design situations that will provide learner opportunities for making choices, for signaling comprehension, for discussing, for reproducing, and so on to minimize the number of questions that must be asked openly.

*Examples of Participation-Level Activities*

- Discussing data presented
- Structured data-gathering activities
- Reviewing data presented in learning centers
- Acting out known situations (dramatic play)
- Walk-through, replication, and verification activities
- Group discussions of a presentation or demonstration
- Opportunities to imitate an observed event

- Reading a story previously introduced or discussed in class
- Using manipulative or hands-on activities
- Counting and quantifying data
- Ordering objects, events, and materials
- Visualizing and then verbalizing or brainstorming consequences or ideas
- Designing simple questions to recall or to translate data
- Modeling or defining behaviors
- Timing activities and programs
- Generating data through recall or book searches

Participation is a dynamic learning level. The teacher's role as catalyst has required careful but enthusiastic guidance of the learner into the experience, since the individual has tentatively accepted this venture into a strange situation—that of incorporating new experiences into one's life. Participation is the level at which the experience is most frequently abandoned or terminated. In some situations that involve choice the termination may be correct. When this is not the case, however, the role of the teacher as catalyst cannot be minimized.

## 3.0 IDENTIFICATION

As noted earlier, completion of this level indicates that the initial learning experience has been achieved. In many situations, however, the learner who has observed and replicated a demonstration (participation) is supposed to have achieved understanding and "fixed" learning. We contend that understanding is not demonstrable until the student moves from the periphery of an experience to an identification with it. Moreover, "fixing" is not required until further along the continuum.

Identification is defined as a union of the learner with what is to be learned in an organizational, emotional, and intellectual context for the purpose of achieving the objective. It is at this taxonomic level that the individual begins to relate overtly self to experience, sees the organization and structure of the idea, gains a deeper insight into its value, and is able to express recognition of achievement. There are four subcategories of identification through which learners move in differing degrees of intensity, depending on the need, drive, or desire created by data collections and explorations.

Reinforcement is that stage of identification that turns miscellany into order. Perceptions, assumptions, and images, heretofore fairly heterogeneous and scattered, are now organized with greater selectivity, thus encouraging attachment to the experience. The decision to identify occurs unconsciously and is frequently unknown to the learner until the next stage. Emotional identification, however, is the process whereby, if the experience is to be retained in memory for any time, it must record itself as a felt thought. The concept of emotion as used here does not have the common connotation of sensation. Emotion here is primarily the nonverbal assimilator of experiential data in which meaning is felt and expressed without words. Following this process, the learner moves to the personal subcategory, or to what is described as an intellectual commitment. This intellectualizing is a more precise verbal phase in which thinking and reasoning are central and are clarified and expressed primarily by verbal thought, teaching the learner to share experiences with others in some symbolic manner.

## IMPLICATIONS FOR TEACHERS, LEARNERS, AND EVALUATORS

The primary strategy of the teacher in aiding identification is to act as a resource leader and moderator prompting students to use data. A constant analysis of the students' learning difficulties and deficiencies is necessary during this phase for the selection of appropriate additional resources and instruction methods. Thus, the teacher will consciously seek to diagnose, to provide corrective feedback to reinforce, and to ask questions that lead the learner to implement, to hypothesize, and to experiment. This will involve further discussions, conferences, field activities, and interaction, as well as measurement of progress.

Throughout the identification process, the learner will experiment by applying, associating, classifying, categorizing, and evaluating data. The learner will be more selective in making assumptions and in confirming hypotheses. There should be greater involvement in activities that are investigative, interpretive, and solution oriented and that result in some symbolic expression of understanding.

Evaluation of this level has two functions. Here a clearer measure of the effectiveness of strategies, prescriptions, and materials is possible. First, the instruments used must be designed to demonstrate not only to the teacher but also to the learner that the learning has been achieved. Measures may be either standardized or teacher-made criteria but evidence must show that the student has accomplished initial achievement. Second, some measure must verify the correctness of the course of learning. This generally involves comparative data based on selected criteria. Depending upon the degrees of sophistication desired, this could involve simple mental checklists or complete studies. It is however, a function that should not be neglected.

### Examples of Identification-Level Activities

- Use of learning and/or reinforcement centers or packages to implement prescriptions or practice skills or content retention
- Conference where individual needs and problems can be dealt with
- Discussion under teacher and/or student direction, in which students exchange points of view and back up statements with accurate data
- Field trips where students can study directly the content of instruction in its functional setting
- Data selection, retrieval and organizing activities
- Independent charting of events in time, history
- Experience where individuals or groups study a particular subject area
- Practical applications of theory through observation, experimentation, and research
- Manipulative lessons that combine psychomotor, cognitive, and affective activities
- Opportunities to publicize or demonstrate learning, such as simple dramatizations, writings, and contests
- Reporting information acquired to class
- Shop work emphasizing skill development
- Conceiving and using provisional assumptions

Identification is an interacting level of which the learner actively participates in the experience using and testing data. The progress level is apparent and has strengthened responses. Learners now do more than stockpile data; they collect resources. Since the learners now risk less, the extent to which they expand their use of these resources and move to the next level suggests to the teacher to-persist in prompting this expansion.

## 4.0 INTERNALIZATION

When an experience touches and continues to influence the life-style of an individual, the highest level of internalization has been achieved. This level is beyond the "half experience" of an intellectual commitment and a learned behavior. It involves a release of the conscious mind. The individual is now viewed as active and self-directive, and progress is no longer controlled from outside. Thus, the experience has been incorporated, is further reinforced, and frequently is a part of unconscious problem solving and creative flashes.

As the experience moves into internalization, there is both an expansion and a fusion. The experience filters into other aspects of the individual's life. Depending on the nature of the experience, it can modify attitudes, beliefs, and philosophy. The experience now has, at least at this time, an uninterrupted access throughout the nervous system and can increase its scope and influence.

When this unconscious influence becomes more than temporary, its new dimensions become pervasive and continuously available according to the current situation. At this level one can see that an experience is not simply mental, physical, or affective. In addition, there is a welding that continuously pulls all the pieces together in a matrix that simply and functionally works toward unity, order, and organization throughout the various levels of experience.

*IMPLICATIONS FOR TEACHERS, LEARNERS, AND EVALUATORS*

Although the teacher may start with simple skill or data rein-forcement and overlearning activities, internalization requires both periodic review and a sensitivity to learners' needs in the level of stimulation and dimensions of the experience. The teacher must now provide situations in which the learner has more control and selection yet resides within progressively broader limits set by the teacher. Problem solving, comparative—contrastive prescriptions, and creative assignments that have roughly designed solutions may be offered initially; but, for the student to proceed to the instrinsic stage open-minded, no-"pat-answers" tasks must be designed with the opportunity built in to discuss developments with others. Probing activities now become not just "probing further" but also probing wider and deeper. Here, in the teaching–learning situation, the learner can exert control either through greater commitment toward and appreciation of the experience or by limiting it to the sharing or expansion stages only. Thus, the implications for the teacher are to reexamine the motivational strategies to be certain that activities include the appropriate learning conditions.

The continuing learner moves beyond seeing simple relationships to generalizing and creating new uses for various aspects of the experience. Concepts are formed, reinforced and/or modified, evaluated to some degree, and transferred to other encounters. Educationally, the learner will be analyzing, transferring, appreciating, inquiring, debating "within and without," and maintaining at first a tentative and then an assured attitude until the situation demands change.

Again, at this category, evaluation takes on a dual responsibility. Which learning is being measured—the retention of the previous level or the expanded and intrinsic changes? The measurement of the first is not difficult and can be accomplished by using different forms of the previous instrument at a time sufficiently in the future to assure some degree of "fixing." The latter is more difficult to measure and, once again, the evaluation must be divided into two parts—that of the alternative teaching–learning strategy selected and that of learner products. Comparative data and criteria for evaluating the selected alternative continue to be important, but the hazard of learner product measurement is that where originality and creativity are concerned, there are no "right" answers, and care in how judgments are to be made must be given adequate forethought. Now, anonymous information-producing instruments are often useful measures for the internal changes.

The implications for the evaluator are that both direct and projective measures be used. Direct measures are those in which the participants demonstrate growth in situations devised by the teacher. They are subject to the control of the participant and may not reveal any more than the image the individual wishes to impart. Projective measures have a greater chance of capturing a true picture of the participant's awareness, values, and beliefs and also present to the teacher areas of concern and misunderstanding for curriculum and program improvement. Thus, although evaluators may include rating scales, checklists, questionnaires, interviews, and so on, they also should make use of projective devices that are open-ended, are anonymous, and provide a rich source of information about the learners and about the teaching strategies that, otherwise, might be missed.

*Examples of Internalization-Level Activities*

- Lessons in which the learner must describe, clarify, and analyze more than one situation, system, or group so as to determine and evaluate similarities and differences
- Discussion or questions that cause the learner to think at higher cognitive levels
- Work experience that provides the student opportunities to use data acquired in one or more practical situations
- Creative situations urging the learners to develop their own styles, whether literary, artistic, mathematic, poetic, scientific, or manipulative
- Seminar activities in which a group of learners bring together research or advanced study findings to resolve problems of mutual interest
- Activities designed to simulate real life situations rather than interpret roles
- Situations requiring originality or synthesizing
- Individual and small group exercises to develop new uses, products, and techniques using selected data
- Study in various fashions of the past to determine future trends
- Panel discussions in which learners present their views
- Probing activities—in-depth studies and "behind the scenes" investigations or research
- Re-creating activities using new media, expressions, point in time, or location
- Role-playing activities providing opportunities to display and express new behavior
- Games in which the emphasis is on strategy, not on skill or chance

## 5.0 DISSEMINATION

At this level the experience extends both inwardly and outwardly with the learner in more control than previously. The teacher's role is to sustain the experience so that the extension may occur. Overlearning activities providing additional scope and familiarization now must be supplemented by a remotivating or differentiated input to assist in both fixing the experience internally and intrinsically transferring it to other areas of learner behavior.

It might be said that the teacher achieves the full experience in moving beyond sharing and self-influencing to an attempt to influence others. The reader will recognize that experience is cyclic as is life. The process that results in a concept formed, a learning achieved, or an experiencing does not stop. The complete act of experiencing is nonending when carried to this level and is responsible for man's progress, good or bad.

An examination of this attempt to influence others separates dissemination into two stages. In the first the individual feels strongly enough about the experience to attempt, through some form of descriptive, personalized communication, to stimulate others to have an equivalent experience. At this stage the experience is seen as beneficial and the individual is motivated to exert the energy to persuade others to have the recommended opportunity. When the focus magnifies, however, to the point at which all or selected others must have the experience, the homiletic stage is reached and can frequently result in continued devotion to the conscious or unconscious search for direct or indirect influence.

### IMPLICATIONS FOR TEACHERS, LEARNERS, AND EVALUATORS

Teaching, at this point, becomes a strategy to provide a variety of vehicles whereby the student can express the experience. The focus now is one of reinstating the experience cycle providing opportunities this time for the student to assume most of the teaching role while the teacher plays only that of critic. Here, attention is given primarily to corrective, supportive, and informational feedback, to the cycle going beyond the existing setting.

As noted previously, the learner now becomes the resource, the presenter, the demonstrator, the motivator, the augmenter, and even the critic of what the experience produces. Behaviors now result in some sort of self-designed product within the structured learning environment and in a self-initiated product outside it. The learner is now reorganizing the accumulated data to meet the set task and to express feelings and ideas. The learner is also comparing and contrasting, this time with an external purpose. Such activities as debating, teaching, selling, and producing demonstrate some form of the experience. Seminars now move from analysis to defense of style, of views, and of values. The learner is now the artist, the craftsman, the professional, the coach, or the leader.

At the dissemination level, the product is the level of communication achieved by the learners. Here the participants are attempting either to lobby or to impress their experience on others. The implication is that the evaluator must be aware of the learner's objective in order to best determine adequate measures of achievement. Evaluation also might consider the length of devotion to the task, to the degree of influence, and to the variety of techniques used. Additionally, the design selected must include provisions for determining how well the individual learner feels the objective has been achieved.

*Examples of Dissemination-Level Activities*

- Simulations in which learners present their case to "juries and to voters"
- Political campaigning—selling a candidate, on issue
- Debates between philosophies
- Seminars structured so that learners must defend their views
- Presentations to illustrate the advantages or excellence of one style, process, view, or value over another
- Demonstration of product designed or created by the learner
- Cross-age and peer teaching and counseling
- Production designed to influence, such as audio- and videotapes, dramatizations, poeting, scriptwriting, publishing, and arts and crafts
- Activities that involve determining audience need for the experience, prescribing for them, and developing strategies for teaching and evaluating progress
- Sales of developed products, productions, and services
- Advertising and recruiting activities

- Writing to influence—letters to the editor, editorials
- Designing and producing own learning center, learning package, minicourse

Dissemination involves primarily a voluntary, outward expression and signals overtly the degree of transfer, of reward, and of motivation achieved by the learner. This expression is a beginning and is cyclical in that it suggests to both learner and teacher the way to new learnings when carefully critiqued.

## SUMMARY

The experiential taxonomy, like all others, will continue to expand with use. It differs, however, in one major respect. It seeks to unify the classification of experience so that teachers can follow one model in developing units, courses, programs, and strategies to achieve the goals in education. It identifies the level at which an objective is initially achieved and offers suggestions for its more complete development at subsequent levels.

The taxonomy appears to refute the theory that seeks to place the teacher only in the role of facilitator. It does this by delineating five separate roles of interaction as the teacher moves through the levels, none of which can be avoided if the learning experience is to reach completion. Teaching strategies based on selected learning principles, which support each of these roles, are provided.

The roles of the learner and of the evaluator are also identified as movement continues through an experience. Initially, the learner merely attends and begins to react to stimuli and then begins to explore as the level of risk increases. Should success seem possible, the learner experiments and extends the experience to a deeper level with the possibility of trying to influence others to have similar experiences. The evaluator, too, progresses from observer to questioner, from assessor to measurer and, finally, to one who determines the degree to which the experience has influenced the learner, and the cycle continues.

In a further review of this chapter, the reader will begin to realize that this taxonomy can be used in a variety of situations within and outside of the field of education. As suggested earlier, its effectiveness in political, economic, and in sociological ations needs to be explained. Whether the several categories are used in a simple walkthrough or in an in-depth study of existing programs or of programs to be developed, may lead to improvement.

# 4 | Total Experience Design for Curriculum Development

The Total Experience Design for Curriculum Development constitutes a basic framework from which instructional systems can be developed. This overall plan precedes the development of instructional programs and allows the creation of many such subsystems. It is developed at a higher level of generalization than is the design for instruction and presents the basic three essentials (program goals and objectives, content, and processes), in such a way that it ensures not only a logical progression through all levels of experience but also, upon completion, achievement of curricular goals as well.

An overview of the use of the experiential taxonomy in curriculum development appears in Figure 4.1. An example of the curriculum format is shown in Figures 4.2 and 4.3.

In basing any instructional program on this design, one must continually consider not only the relationship between the goals, the

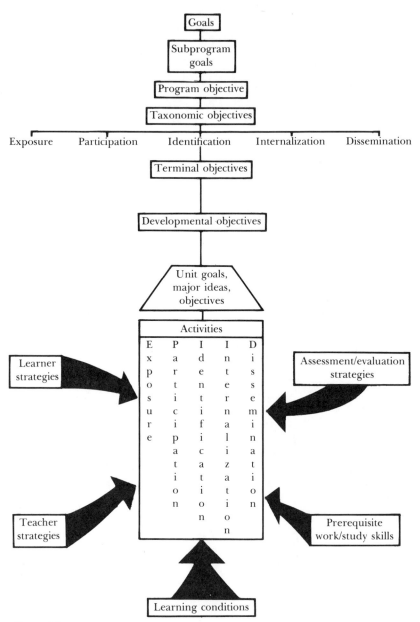

**Figure 4.1**
Total experience design for curriculum development: An overview of the entire development process.

*THE EXPERIENTIAL TAXONOMY*

## Unit goal

Students will recognize that the exchange of goods and services is essential to human development.

## Rationale

Interdependence and trade are facts of modern economic life. Students should realize no country is self-sufficient in resources.

## Major Generalizations

Trade relations are marked by custom, law, and economic agreements.
There are similarities and differences in the way people exchange goods and services.

## Concepts to be taught

| | | |
|---|---|---|
| Economic interdependence | Service | Export | Barter | Money |
| Customs | Trade | Society | Credit | |
| Laws | Import | Exchange | Goods | |

## Objective(s)

The student will demonstrate understanding of similarities and differences in the way trade is conducted in selected cultures: Polynesia, Japan, United States.

**Figure 4.2**  Sample unit format: Introductory steps.

39

content, and the processes but also the three levels of decision making: the program, the learner group, and the individual learner. No curriculum has lasting substance unless these three levels are considered throughout its entire development.

At the program level, the curriculum design is concerned with the overall goals and with several levels of objectives. Statements establishing the need or justification for the program often precede these goals. The goals themselves state in general terms what the curriculum is designed to teach. Subgoals further expand the program goal and clarify areas of focus. A simple illustration of these is found in Table 4.1. In developing the program objective the planner explains what the student or learner will demonstrate upon completion of the entire curriculum, whether it be an understanding, an appreciation, or a behavior.

Up to this point our framework has followed the normal steps for curriculum construction. It is here that it begins to differ. Once the program objective has been determined, the Total Experience Design requires that taxonomic objectives be written. The intent of taxonomic objectives is to specify how the program objective will be achieved through exposure to, participation in, identification with, internalization of, and dissemination of the major concepts, aspects,

**Table 4.1**
Sample Statement of Curriculum Goals

---

- The varying social, scientific, and technological forces bombarding the world today provide overwhelming evidence of the need for economic literacy on the part of every citizen in our nation. It is to meet this need that the present program has been developed.
- The program for economic literacy has been designed to enable and encourage all students in the school district to advance in their understanding of, and to achieve a high degree of literacy in the field of economics and its effect on their lives and on the lives of their nation's citizenry. It is the purpose of this program to assist students in
  - Understanding and appreciating this nation's system of capitalism and free enterprise
  - Acquiring an understanding of their role in the economic system of their country
  - Developing an appreciation for the economic and social achievement made possible by the operation of a free people in a free society
  - Understanding that, as in other areas, continuing change presents new economic problems for resolution
  - Developing an attitude of responsibility to participate actively in the solution of these problems

---

and activities of the program. By specifying to which components the learners will be exposed, in which types of activities they will participate, with which concepts or generalizations they should identify, which understandings they should internalize, and which products they should disseminate, the planner presents in taxonomic form the relationship between the objectives, the content, and the processes, shown in Table 4.2.

Once these are completed, terminal performance objectives that further delineate what learners will be able to do or what new behavior they will manifest, can more easily be determined. The primary purpose of this level of objective is to designate for the unit planner and for the teacher which understandings, which appreciations, and which behaviors are included. The reader will note that the program objective in Table 4.2 states that these will be demonstrated for "the economic system as it exists in the United States today" but does not identify them. Several that might be included are shown in Table 4.3.

The objectives described have been all-encompassing, broadly describing the scope of the program. They cover the life of the program that could extend from primary through the secondary or college levels. The curriculum design must provide an ordering process of achieving them through developmental objectives (Table 4.4). Here one thinks in terms of planning and ordering experiences along

**Table 4.2**
Sample Statement of Unit Objectives

---

Program Objective
Upon completion of the economic literacy program, the student will demonstrate an understanding about, an appreciation of, and a responsibility for the economic system as it exists in the United States today.
Taxonomic Objective
1. The student will be exposed to the concepts, achievements, and conflicts found in our economic system today.
2. The student will participate in selected activities examining the development of economic concepts, achievements, and conflicts present in the United States today.
3. The student will identify with the responsibilities inherent in those concepts, achievements, and conflicts existing today in our nation's economic system.
4. The student will internalize an understanding of and a need to participate in those concepts, achievements, and conflicts found in our economic system today.
5. The student will disseminate an appreciation of the concepts, achievements, and conflicts existing in our economic system today.

---

**Table 4.3**
Sample of Unit Terminal Performance Objectives

---

Through a series of taxonomic experiences, learners will
1. understand and appreciate the interrelationship between the allocation of resources, economic growth, productive resources, technological progress, the role of government, standards of living, specialization, and exchange between nations
2. be able to examine decision-making processes at the local, state, and national levels as these relate to the economic system
3. identify and appreciate the economic achievements of our system of capitalism and free enterprise
4. identify those economic conflicts needing resolution, and
5. be prepared to participate in the resolution of economic problems arising during their lifetimes.

---

a sequence such as introduction, reinforcement, and mastery, or Levels 1, 2, and 3. This requires that the curriculum planner arrange the objectives in such a way as to provide a logical scope and sequence. The curriculum writer now decides when to introduce, reinforce, or add to the objectives, and in what order they are expected to be mastered. As the sequence is developed, specific units or learning sequences become relevant, and ideas for experiences for a particular grade, age, or group are developed. When the scope and sequence have been determined, specific planning through unit development for individual and group experiences can begin. At this point a step-by-step total experience design becomes pertinent. A step-by-step guide for developing curricular experiences takes on real mean-

**Table 4.4**
Sample of Learner Developmental Objectives

---

Level 1
  A. Learners demonstrate an understanding of economic concepts as they relate to their lives in the home, the school, and the community.
  B.
Level 2
  A. Learners recognize the technological progress achieved within our system of capitalism and free enterprise.
  B.
Level 3
  A. Learners can generalize relationships between economic growth, resources, and standards of living.
  B.

---

*THE EXPERIENTIAL TAXONOMY*

ing only when the scope and sequence for the particular course of study are determined.

## A STEP-BY-STEP PLAN

The recommended plan presented here is based on the experiential taxonomy, which serves as an organizing format both for the total curriculum and for the units of work to be developed to implement the plan. The objective of this step-by-step plan is to present a taxonomically based approach to help teachers plan effective classroom experiences. Not only can this model be the basis for building an effective curriculum, but it can also help teachers understand better the teaching–learning act. Figure 4.2 supplies a sample unit format for the introductory steps.

### STEP 1. DEVELOPING A RATIONALE AND GOALS FOR THE UNIT

Today, more than ever, with the demands made upon the schools to include what so many different agencies, interest groups, and individuals feel are musts for each student, curriculum planners need a defendable rationale for what is selected and what is rejected. This justification must begin with goals but must extend through content and processes as well. To be certain that the rationale or reason for selection is clearly understood, each unit should have a raison d'être that relates to its function in the classroom as well as to the developmental and terminal objectives that it seeks to achieve.

The curriculum planner should start with the basic questions: Why should the unit be taught? Why is it important? This focuses the rationale and allows the listing of goals not just for content but also for skills and processes. These goals are statements about what the unit is expected to accomplish. They are broad statements that tend to be all-inclusive rather than explicit. It is not unlikely that a goal encompasses and implies several generalizations, relationships, concepts, and big ideas. Goals are those aspirations that the learner is expected to remember or to internalize and, therefore, to act upon long after the unit is taught. One need not specify concepts, generalizations, or big ideas at this point since the next task in the sequence is to delineate those appropriate to the unit.

It is initially important that goals be written as a point of continued reference throughout unit development. To avoid a lack of focus that can easily occur and can destroy the effectiveness of the unit, the curriculum planner needs to reexamine the unit continually with regard to these goals at each step.

*STEP 2.  DELINEATING MAJOR*
*GENERALIZATIONS AND CONCEPTS*
*IMPLICIT WITHIN THE GOALS*

Every goal has implicit within it some "big ideas," some major generalizations and some concepts that are important in the learning sequence and that need to be recognized and made explicit. These generalizations and concepts, although summarized and implied in the goals, are central ideas that curriculum planners and teachers want and expect learners to internalize and act upon. For instance, if the unit goal were, "The learner will understand the effect of environment on transportation in major world cultures," the major ideas selected could be

- Transportation forms and usage developed from survival needs and available resources.
- There are similarities and differences in transportation forms today in different world cultures.
- Competition between world powers brought forth newer, as well as a sophistication of older, transportation forms and usage.
- New forms of transportation brought about changing values, which increased demands for further change.

Concepts chosen as central to achievement goals could be those of survival, world power, and resources. Other main ideas centering around trade, careers, and political and educational pressures could be added. It is at this step, however, that the major generalizations and/or concepts to be learned should be written out and listed. Planners should be certain that these are, indeed, those major elements that learners need and can gain from the curriculum.

Before proceeding, a clarification should be made between generalizations and concepts. In unit development concepts involve abstract ideas, thoughts, and associations. They are frequently limited as to time, to space, and to usage. Generalizations cover those conceptions, principles, or inferences that extend more widely—indefinitely or universally. Thus, concepts are the particulars from which generalizations are formed. Planners need to identify in advance which of these the unit is to develop. This examination frequently results in modifying the original unit goal to one that is more credible to the teacher implementing the unit.

### STEP 3.  LISTING UNIT OBJECTIVES SEQUENTIALLY AND IN BEHAVIORAL TERMS

Each unit in preparation and such planned learning sequence needs a listing of behaviorally written objectives. Once the major generalizations, concepts, and big ideas have been identified, the planner can write objectives for these and decide upon an appropriate order of presentation. The objectives are written initially without a particular sequence. Upon completion, the planner orders them in a manner appropriate to the learning situation. The following objectives might be written for the major ideas mentioned previously.

Through a study of three different world cultures, the learner will

- identify how forms and uses of transportation developed
- recognize similarities and differences found in present-day transportation forms and usage
- identify and support by the activity selected four common pressures for change that resulted in present-day transportation forms and usage

Ordering objectives is important, since it enables the planner to determine not only the sequence but the time frame necessary. Furthermore, it frequently results in a realization that there are subunits that must be prioritized and taken through the taxonomy before the unit goal is achieved. Ordering also causes the planner to decide whether an objective should be placed at a different developmental level so that its attainment would be a prerequisite to the unit being designed. The preceding objectives are an excellent example of this.

Objectives should be written in behavioral terms and should describe specifically some skill, attitude, or behavior that students are expected to internalize. It is important to remember that the achievement of a skill, attitude, or behavior should be measurable.

When possible, the unit should include the evaluation procedure for the objective. This most frequently occurs in the skill areas where one standard is set. When there is more emphasis on higher levels of thinking and on individual interests and learning styles, the objective should suggest direction and limits, allowing for some teacher–learner discretion.

*STEP 4.  IDENTIFYING PREREQUISITE LEARNER
SKILLS NEEDED INITIALLY AND AS THE
UNIT DEVELOPS*

A great many units fail because of inadequate attention to the skills necessary to begin the unit. Pinpointing these skills is really a two-part process. First, those skills of content, attitude, and physiological development that the learner must possess in order to effectively begin the unit or learning sequence must be identified and teaching started so that the user is well aware of them in advance. The teacher can then assess the level of student or class readiness. If the skills of a unit are too advanced, learners will express little interest and cannot be expected to make much progress. This important step is crucial to learner success. After level skills have been identified, planning in terms of the specific grade, the maturity, and the achievement levels appropriate to the unit can begin. Upon completion of this step, the planner should have firmly in mind the kinds of learners for which the unit is being prepared. Then, if the unit or learning sequence requires additional special skills, activities, or achievement expectations as it progresses, these are noted to be dealt with as the unit is taught. Frequently, these can be taught within the unit context. In other instances, however, particularly where these special skills, activities, or achievement expectations may cross disciplines, they are dealt with in other classes or by other teachers. Planners should caution teachers to deal with this problem and to plan so that learners can be taught these skills.

*STEP 5.  PLACING ACTIVITIES FOR OBJECTIVE
ACHIEVEMENT IN SEQUENTIAL ORDER*

This step and Step 6 are considered key steps in developing a taxonomically based unit and should be considered inseparable. After the objectives and prerequisite learner skills have been identified, activities, some more pertinent than others, are planned and ordered in the most appropriate sequence for objective achievement. At this point, the planner needs to consider all possible contingencies such as available resources, appropriate learning principles, necessary teacher strategies, and learning behaviors, in making decisions. Figure 4.3 supplies a sample format for these steps. Initially, activities should be sequenced as appropriately as possible in an anticipated order. Care should be taken to ensure flexibility in the activities so that implementers can deal with such other factors as individual learning styles, interests, and rates. The planner is really providing an ordered sequence of those anticipated activities a learner can follow in order to achieve the stated objectives, after considering as many variables as possible. These activities invite the student to learn and lead to interacting with and expanding the experience, with the resultant expression in the behavior of the student. It must be mentioned that no curriculum planner can make definitive the total direction of the unit. The activities are designed primarily as a road map for the professional and should not be wedded to one route only.

*STEP 6. KEYING ACTIVITIES TO THE*
*EXPERIENTIAL TAXONOMY*

After initially sequencing the activities, the planner should key these to the major taxonomic categories. It is here that planners could recognize two possible errors in unit construction: omission and improper sequencing. Should activities be missing for any taxonomical category, the constraint on unit implementation and objective achievement becomes obvious, and the planner can make the necessary corrections. When activities are found out of taxonomic sequence, adjustments can now be made to ensure an orderly learning process. Furthermore, it is at this step that the unit developer can see the necessity for additional activities to strengthen learning at the taxonomic level. As the activities are keyed to the taxonomy, planners should be aware of the role models of each person involved in the teaching–learning process. Since these are described in Chapter 6, they are only listed here. See Table 4.5.

**Table 4.5**
experimenter

| Taxonomic level | Learner | Teacher | Evaluator |
|---|---|---|---|
| exposure | attender | motivator | observer |
| participation | explorer | catalyst | questioner |
| identification | experimenter | moderator | assessor |
| internalization | extender | sustainer | measurer |
| dissemination | influencer | critiquor | determiner |

**Figure 4.3**  Curriculum unit format: Unit development chart.

Objective: The students will demonstrate understanding of similarities and differences in the way trade is conducted in selected cultures: Polynesia, Japan, and United States.

| Taxonomic level | Activities | Learning principles | Learner strategies | Teaching strategies | Resources | Assessment |
|---|---|---|---|---|---|---|
| Exposure | 1.1 Present information about at least two specific cultures. 1.2 Explore interest centers about cultures presented. 1.3 Introduce and discuss concepts of trade, economics and interdependence. 1.4 Discuss and define terms such as goods, services, import, export, factor, etc., as they relate to interest centers and concepts introduced. | Extrinsic motivation Focusing Anxiety level | Seeing Hearing Listening Recognizing Responding | Incentive conditioning Data presentation Directed observation Data exploration | District Resource Persons Directory | Observing student reaction. Student response to discussion and interest centers. Student readiness for participation. |
| Participation | 2.1 Students bring items to add to interest centers. 2.2 Discuss those items as they relate to trade using a class retrieval chart. 2.3 Use resource materials to clarify concepts (text and A.V. material). 2.4 Complete class retrieval chart. 2.5 View film "Why Communities Trade Goods" to determine how goods are exchanged in cultures selected and how goods and services move from culture to culture. | Initial guidance Meaning Exploration Chance for success. | Observing Exploring Reacting Discussing Deciding | Modeling–recall Expanding data bases Manipulative and tactile activities Ordering | Text: Concepts and Values Grade 4, pp. 144–200 Four to 6 pictures of each culture Study prints, "Living in Japan" Film: "Polynesian Culture" | Student additions to learning centers. Student understanding of concepts through questions and discussion. Completed retrieval chart. Appropriate student decision. |
| Identification | 3.1 Use films and other available resource materials to collect data about trade with and between cultures using individual retrieval charts. 3.2 Discuss the data collected identifying similarities and differences in exchange and transport of trade goods. 3.3 Organize information into categories of similarities and differences. 3.4 Share and discuss information. | Personal interaction Knowledge of results Reinforcement | Investigating Classifying Categorizing Explaining | Field activities using data Discussion–Conferencing interaction Hypothesizing Testing | Textbooks: Addison-Wesley: Regions Around the World. Saba: Life in Osaka Prefecture. Geis, Darlene: Let's Travel in the South Seas. Siedler: Japan, Hawaii, Midwest, and Great Plains | |

| Taxonomic Level | Activities | Learning principles | Learner strategies | Teaching strategies | Resources | Assessment |
|---|---|---|---|---|---|---|
| Internalization | 4.1 Discuss application of the information to imaginary countries.<br>4.2 Divide into groups, each to represent an imaginary country.<br>4.3 Provide the following information about each group's country: natural resources, climate, geography, technology, cultural beliefs, social and political organizations.<br>4.4 Groups decide what each country needs to import and export.<br>4.5 Develop a report on the economic needs and trade relations of the country in an approved format. | Overlearning<br>Intrinsic transfer<br>Differentiated input | Analyzing<br>Inferring<br>Comparing<br>Contrasting<br>Organizing | Skill reinforcement<br>Re-creation<br>Comparative-Contrastive analysis<br>Summarization | | Student decisions.<br>Group interaction.<br>Group conclusion.<br>Group written report. |
| Dissemination | 5.1 Groups report to class on findings.<br>5.2 Prepare a prospectus to convince investors to come to their country.<br>5.3 Prepare a bulletin board on the material gathered. | Extrinsic transfer<br>Reward<br>Intrinsic motivation | Presenting<br>Communicating<br>Influencing | Reporting<br>Oral Presentation<br>Group dynamics | | Peer interaction and evaluation.<br>Completed prospectus.<br>Completed bulletin board. |

Comments:

This objective provides opportunities for students to explore and demonstrate understanding of important aspects of economies. At the exposure level, one can utilize community resources for the presentation about cultures. It should also be noted that the cultures selected should represent an industrialized culture and a developing culture. The transition from known data about specific cultures to imaginary ones is crucial to the achievement of the objective. Be sure you have enough data to present to the groups. A major task for the teacher is the gathering of this information. You could use information from existing nations, but do not give students name of country.

## STEP 7.  IDENTIFYING LEARNING PRINCIPLES
##          PERTINENT TO EACH OBJECTIVE

Behind every successful learning experience are compatible and appropriate principles of learning. Some of the most obvious used in unit construction are motivation, focusing, anxiety level, initial guidance, meaning, chance for success, personal interaction, knowledge of results, reinforcement, overlearning, transfer, differentiated input, and reward. These learning principles thread throughout the entire taxonomy but do have greater effectiveness at particular levels in an experience continuum. Motivation, for example, is needed in every phase but is the dominant teaching mode at the exposure level. As unit developers work with the taxonomy, they will find this true of most of the principles mentioned. It is at this step that planners should reexamine the selected activities in relation to the learning principles, to determine whether or not they truly utilize these basic principles. We believe that the selection of sound learning principles is basic to successful classroom experiences for learners and that this must be done at the time of unit development. Those principles chosen as best supporting each taxonomic level of unit development are noted in Table 4.6 and are described in the following chapter.

**Table 4.6**
Taxonomic Sequence of Learning Principles

| Exposure | Participation |
|---|---|
| Extrinsic motivation | Initial guidance |
| Focusing | Meaning exploration |
| Anxiety level | Chance for success |
| Identification | Internalization |
| Personal interaction | Overlearning |
| Knowledge of results | Intrinsic transfer |
| Reinforcement | Differentiated input |

Dissemination
Extrinsic transfer
Reward
Intrinsic motivation

## STEP 8.  LISTING OF EXPECTED LEARNER
##          BEHAVIORS

Learner behaviors are those actions, strategies, and skills that the student actually learns, does, and/or practices in the performance of an activity. To ensure that they visualize all aspects of unit development, planners must seek to identify those that learners will most frequently be called upon to perform at each taxonomic level. This is a crucial step since it provides additional information about other student needs for the teacher, for the evaluator, and for the planner. Furthermore, it focuses teacher attention on which behaviors are appropriate and when for the activities chosen. In evaluation, it provides additional checkpoints in the selection or the creation of strategies and of instruments. For the planner this simple act of visualizing and listing what the learner is to do for each activity enables the unit creator to refine or to improve decisions made previously in activity selection and sequencing. This step is dealt with at this particular point, because we believe that a clearer picture can be visualized after consideration of those learning principles identified with each activity (see Table 4.5).

Learner behaviors are innumerable and are found at the various levels of the taxonomy. To better understand how these might be selected for the five major taxonomic levels, the reader should examine the following samples.

EXPOSURE: Seeing, hearing, smelling, tasting, touching, noticing, reacting, recognizing

PARTICIPATION: Observing, discussing, exploring, assessing, visualizing, directed reading, estimating, manipulating, ordering, collecting, modeling, defining, listening

IDENTIFICATION: Associating, classifying or categorizing, explaining, experimenting, mapping, evaluating, hypothesizing, interpreting, investigating, writing or drawing, applying, charting, reading or observing for information

INTERNALIZATION: Analyzing (deducing, inferring), summarizing, generalizing, comparing or contrasting, inquiring, in-depth probing, transferring

DISSEMINATION: Communicating, debating, influencing, demonstrating, presenting, motivating

## STEP 9. LISTING APPROPRIATE TEACHING
## STRATEGIES FOR EACH ACTIVITY

It is our contention that for each activity carried out within a unit of work the teacher has a definite role (see Table 4.5). The strategies for implementing each of these roles, however, are not as static and are often as different as are the activities. For example, activities listed at the exposure category of the taxonomy require the teacher to assume the role of motivator. The strategy chosen by the teacher may involve simply asking naive questions to arouse interest, using audiovisual material, demonstrating a skill, or arranging a tour of selected facilities. For a total experience unit design to be successful, its planner must be able to visualize what it is that the teacher must do within each role along the taxonomic sequence and must list these along with the unit. The teacher who implements the unit as written then has a clear idea of the planner's intent at the time the unit was developed and, hopefully, piloted. Although there are a wide variety of teaching strategies, some that we have used in unit development are listed in the following and fall within the teaching modes presented in Chapter 8.

EXPOSURE: Defining, lecturing, data presentation, using audiovisuals, provoking attention, directing observation, providing interest centers, promoting problem identification, demonstrations

PARTICIPATION: Diagnosing or prescribing, signposting, clarifying discussions, providing supportive feedback, promoting or implementing ordering, dramatic play, modeling or imitation, manipulative activities, learning at centers, recall or replay activities, simple data-gathering lessons

IDENTIFICATION: Continuing diagnosis, prescription, and treatment to meet individual or common needs; providing corrective feedback; promoting additional resources; testing, questioning, or conferencing; promoting hypothesizing, experimenting, simulation, and drill activities; encouraging group discussions; use of reinforcement centers, field experiences; peer teaching

INTERNALIZATION: Remotivating, structuring situations for problem solving, analysis of more than one system, role playing, creating, expanded skill practice, encouraging student probing.

DISSEMINATION: Student expressions (reporting, oral presentations); productions; dramatizations; seminars; the supportive, informational, and corrective feedback necessary

## STEP 10.  LISTING EVALUATION AND ASSESSMENT TECHNIQUES

For each objective, teachers need verification that the students have functionally learned the skills, concepts, and content to achieve unit objectives. At this step in unit development, planners need to list at least one method in which teachers can check the learning achievement and skill accomplishment of the students (see Table 4.5). They must do this for each category within the unit. Often evaluation and assessment techniques are simple and involve little more than spot checking. In other instances, evaluation may require a quiz, a test, a practical application, or planned opportunities for demonstrating behavior. If they have carefully written objectives and activities, planners can easily select appropriate techniques. They should choose evaluation strategies that best help the teacher to know whether the student has successfully mastered the objective. The following suggestions as well as the later chapter on evaluating the teaching–learning act will help a curriculum writer.

EXPOSURE: Observing learner reaction to the initial activities to determine attention; understanding of terms, scenes, and purpose; and readiness and/or willingness to proceed

PARTICIPATION: Examining student choices; signals of understanding or of lack of understanding; replications; discussions; questioning to determine understanding; ability to succeed; and, where appropriate, explanation of how the learner "would do it" if given the opportunity

IDENTIFICATION: Using criteria, teacher-developed tests or assignments, and mental or actual checklists to assess student progress and teaching or unit effectiveness

INTERNALIZATION: Using projective measures such as open-ended, anonymous response questionnaires and/or direct measures such as rating scales and interviews; using a post- and retest method in which a different test form or assignment is given at a later date and is compared with the original test or assignment to determine retention

DISSEMINATION: Using student self-evaluation instruments; evaluating the time devoted to tasks, the variety of techniques employed to use or to promote the learning, and/or the degree of influence achieved

## STEP 11. CITING NECESSARY RESOURCES FOR ACTIVITY AND OBJECTIVE ACHIEVEMENT

No unit is complete without a citation of appropriate resources that may be necessary for objective achievement. The reason for withholding this step until this point is simply that it is assumed that the planner will be collecting these at every step along the way and now can cite the ones most useful to the teacher. No attempt, however, should be made to limit the user to those mentioned. The purpose of listing resources is to provide the user with appropriate examples that, if not available, will provide a point of reference for substitution. Planners with only one district in mind should first note what is available for the teachers in that particular district. They should be careful, however, to select from the total spectrum of resources available rather than to limit teachers to one district library, to one media center, or to one audiovisual catalog. Whenever possible, they should critique these additional sources prior to listing or should note that selection was made on the basis of professional reviews or of other techniques. In addition, should planners refer to community resources, the persons involved should be briefed regarding their role and its place in the total unit.

The Total Experience Design for Curriculum Development as shown in Figure 4.1 ends with Step 11. Many planners and teachers, however, are rightfully concerned with the relationship of their units to creativity, to critical thinking, and to problem solving. Although these are generally included in the original steps, planners desiring to emphasize them in unit development should refer to the taxonomic sequence provided in Chapter 7. The five taxonomic steps for creativity are *exposure*—motive to produce; *participation*—visualization of production; *identification*—experimenting with the idea; *Internalization*—completing a product; and *dissemination*—admiring, showing, and sharing. Those listed for critical thinking are *exposure*—recognizing the variable; *participation*—data collection and variable definition; *identification*—organizing and structuring; *internalization*—generalizing; and *dissemination*—using the variable. The taxonomic steps for problem solving are *exposure*—problem identification; *participation*—data gathering; *identification*—testing an optional solution; *internalization*—choosing a particular solution; and *dissemination*—implementing and influencing.

## FINALIZING THE CURRICULUM PACKAGE

For maximum effect, the preceding steps should be noted in whichever unit format is used. This will ensure that users are aware of these elements and encourage their use in the curriculum. After the steps are completed, the unit is ready to be put into a functional format for curriculum development. The format cited has proven to be functional for teacher use, but variations are always possible. It should be used as planners make notes, write sequences, and prepare units. The format also serves an organizational function in that it is a way for planners to recheck their work, to make amendments, and, generally, to prepare the unit for classroom use. One element that teachers who have used this format have appreciated is the space at the bottom of the page for comments. These comments should have specific emphasis, explaining material and its rationale and bring the components into a more understandable whole.

A truly complete and professional curriculum package that brings together theory and practice now exists. This format is unique in curriculum development, yet it allows for expansion to meet the individual situation. In it the planners have focused on the continuing teaching–learning act in terms of a learning sequence and have, in addition, identified elements of learning theory and of the structure of experience. A unit carefully following the steps listed should result in a strong learning package and should increase teacher effectiveness in the classroom.

One aspect of the teaching–learning act not touched upon specifically in unit development was how to deal with individual learners. The individual learner is the third dimension of decision making mentioned initially in this chapter. Once the unit has been completed, following the preceding steps, the planner must try to include provisions for meeting the needs of the individual learner within the classroom. Thus, a review of the unit in its initial form should result in consideration of alternative activities for

- specific and identified learning styles such as visual, auditory, and kinesthetic
- differing levels of interest and background, which should be listed and identified according to heterogeneous grouping
- opportunities for gifted and/or slow learners
- individual participation in committees, in subgroups, in special projects, and in research, and individual development

Consideration means making explicit these individual student needs. Without this dimension even the most carefully planned unit has limitations.

Another problem in terms of individualization is to distinguish between learning principles and learning processes, in the curriculum planning. A lengthy discussion of this problem is not the central issue of this chapter, but at this point a brief statement is pertinent. Learning principles are those underlying components of learning that remain constant throughout any learning experience. Learning processes are those internal activities each student goes through in achieving new sets of learning objectives. Learning processes are contingent on individual learning styles, modes, and preferences.

Students having different learning modes require different learning principles. For example, a student whose learning mode is auditory will respond to certain elements of a learning principle like motivation differently than will a classmate who has a visual learning mode. The learning process through which students go will vary as previously noted, yet the learning principle will be the same. In using a learning principle such as motivation, the teacher will select for youngsters different kinds of stimuli and different strategies of motivation, with the expectation that individual students will achieve the planned objective. In essence, the principles and the product (objective) remain the same; only the process varies. The instructional problem is, of course, to plan in any learning sequence a varying method or strategy to meet those individual needs. The key to the solution of this problem is the ability to identify varying learning styles and modes. This involves three elements. First, the teacher must be familiar with the various learning styles and modes; second, the teacher must know the students who participate in the learning experience; third, the teacher must become familiar with all resources, especially those emphasizing each particular learning modality. Should the teacher accomplish all of these, there will be little difficulty in individualizing an instructional unit to meet the needs of students with varying learning styles and modes.

## SUMMARY

The Total Experience Design follows a logical sequence of objectives and events that assure the user of the developed curriculum that the activities suggested are based on identified principles of learning and have been envisaged from the viewpoint of the learner, of the teacher, and of the evaluator, and that the materials most useful to the implementer have been included. In addition, by including the experiential taxonomy, the design provides a clearer picture as to how the program goals will be achieved and the units developed. The units themselves are now created with attention on both product and process evaluation, which enables the teacher to see more clearly the level of achievement reached both by the class and by individuals.

# 5 | Learning Principles: A Taxonomic Organization

Learning principles are definitional representations of those processes that provide a philosophical and psychological frame of reference for learner activities and for the acquisition of learner skills. They provide also a means for planning teacher strategies, learner skill expectancies, and anticipated learner strategies. Every teacher needs to be aware of these principles and the need to use them when planning, implementing, and evaluating curricula.

Although one may identify many principles for effective taxonomic planning, a functional grouping is appropriate. We have identified 15 learning principles as basic to the pedagogical process and functional to experiential taxonomic planning. They are extrinsic motivation, anxiety level, focusing, initial guidance, chance for success, meaning–exploration, personal interaction, knowledge of results, reinforcement, intrinsic transfer, overlearning, differentiated input, extrinsic transfer, reward, and intrinsic motivation. Each of these can be used hierarchically in that there is a logical sequence to them as one thinks about learning. In dealing with this pattern, we will briefly define and key each learning principle to the major categories of the experiential taxonomy.

It should be noted that, though these learning principles are taxonomically sequenced, as basic principles of learning they can be present at more than one taxonomic level. Furthermore, the definitions presented are ours and may or may not readily conform to any particular school of thought or existing theory. Each principle has been placed somewhat arbitrarily in a particular taxonomic category because it is at that level that the learning principle can be most appropriately introduced, that it can be most effectively emphasized by the teacher, or that it is considered fundamental and necessary. This is not to say that it cannot occur at succeeding or preceding taxonomic levels in order to meet student needs.

As previously noted, a learning principle occurs at all taxonomic levels but is fundamental at one category. For example, motivation is used at all categories but is a fundamental and necessary principle at the exposure level. If motivation does not occur at this level, the experience will not progress beyond that point. It must also be understood that in any group or classroom experience, different students will be at different taxonomic levels and, thus, the teacher may need to build strategies that relate to more than one of the learning principles and to more than one level of the taxonomy. We hope that this taxonomic organization will be helpful to teachers as they plan effective classroom experiences for their students through the Total Experience Design discussed in Chapter 4. It should also be noted that these learning principles exist in a number of learning situations and involve such aspects as student learning styles and modes, materials and media, the socioeconomic status of students, established teacher–student relationships, peer relationships, and personal qualities of the students. Furthermore, it should be remembered that the role of the teacher changes as the student moves through the taxonomy and as learning takes place. We mentioned in an earlier chapter that we do not regard the teacher as merely a manager or a facilitator. The teacher is always teaching and fills five distinct roles in the taxonomic sequence. They are motivator (exposure), catalyst (participation), moderator (identification), sustainer (internalization), and critiquor (dissemination) (see Chapter 6).

Following is a brief definition of each learning principle at the taxonomic level in which it is emphasized or fundamental.

## EXPOSURE

### EXTRINSIC MOTIVATION

This learning principle deals with the way in which the student is exposed to and reacts to a learning experience. The student can be motivated in many ways, but initially this involves the senses. According to this principle, the learner should have some freedom to interact with the teacher-planned stimuli so that there will be a readiness to continue the experience. Motivation involves a whole spectrum of teaching strategies and the teacher should select those that prove effective with students. These, of course, will change as situations and students change. At the exposure level of the teaching process, the teacher treats motivation as an extrinsic learning principle to be certain that it involves stimuli from outside the past experience of the learner and that it requires some initial extrinsic manipulation of the environment by the teacher. Motivation remains a constant part of the teacher–learning process. As it moves taxonomically through an experience, it becomes progressively more intrinsic. The major shift from extrinsic motivation to intrinsic motivation begins at the internalization category of the taxonomy. The intrinsic stage, however, is strongest and most dominant at the dissemination level. Motivation at the exposure level literally means creating a drive, a felt need, or a desire to learn. Its characteristics are alertness (attention) in the learner, a desire to achieve or to be able to do something, a feeling of possible success, and a feeling of enjoyment, reward, or purpose. Teacher techniques for motivation include manipulating the variables most responsive to the situation. These variables fall into three classes: (*a*) the individual who does the learning; (*b*) the nature of the task; and (*c*) the conditions under which the particular learning occurs. The conditions are defined as elements that teachers can control, change, or vary. The following are examples:

- Learner tension or concern—when it exists to a moderate degree it assists in motivation; when it is extensive, it diverts learner energy to dealing with the tension and not with the task.

- Learner reaction to stimuli—as pleasant or unpleasant. In moderation this may increase motivation; in excess, it debilitates; in absence, it decreases motivation.
- Learner interest—as curiosity; the more curiosity, the greater the learning accomplished.
- A feeling of victory or success—this greatly enhances motivation. When the student experiences success with the right degree of difficulty, motivation increases. Lack of success or a too difficult task decreases motivation.
- Knowledge of results—the more specific, positive, and supportive the feedback, the greater the increase of motivation.
- Relation of activity to reward—where the activity itself is rewarding, intrinsic motivation is produced and motivation increases. Extrinsic motivation is dependent on and changes with the situation.

The functions of motivation are to energize and activate the learner, to direct the variable and persistent activity of the learner, and to emphasize the consequences of response.

*FOCUSING*

Concomitant with motivation is focusing. No motivational stimuli have any meaning or make any sense unless there is a learner response. Focusing involves directing the learner response to the sensory exposure involved in motivation. Here the job of the teacher is to direct the learner's attention, through a variety of strategies and techniques, to those elements of the planned experience that are most pertinent to the expected outcomes. The learner needs to attend to data or material to gain a sense of their meaning and for enough success to respond positively to the exposure activities. The learner thus establishes a readiness through additional data exploration to move to the participation level of an experience. Focusing of the learner's attention on specific elements of an experience can also prevent the broadening of the experience beyond the bounds of realizable achievement. The discerning teacher will, however, plan the focusing motivation carefully and develop a range of stimuli or a variety of activities so that a direct focusing on specific elements of the experience will create the appropriate response. When planning the focusing activities, the teacher should remember that focusing also depends on the developmental and maturational level of the learner. Human beings fortunately are both curious and unique. Once a person is motivated, attention can be focused. The teacher can present specific data, demonstrate a principle, or show how something works. Focusing also involves directed observation when the teacher directs attention to particular stimuli and begins to establish parameters for the observation or experience. The teacher needs, however, to consider the uniqueness of class members. The teacher can also direct the learner through motivational focusing to begin to interact with selected data or with specific stimuli and in this way to establish a readiness for further experience.

## ANXIETY LEVEL

As the learner is led to attend more specifically, the realization that there is an external purpose can cause either a positive or a negative reaction. This degree of feeling regarding a learning situation derives from the pleasure–pain principle and, if not considered, can cause an immediate rejection of further progress. Anxiety can be positive or negative, and in the learning situation it may exist in varying amounts. If the experience appears too easy, it will be rejected as early as if it were too difficult. And, although both extremes appear to result in behavior difficulties, neutral feeling is ineffectual. Thus, the teacher must be certain, as the purpose becomes clear, that, for the learner to make voluntary progress, mild anxiety is not necessarily an undesirable phenomenon. It does have a function that should be considered. With simple material, learning speed is frequently greater when the anxiety level is moderately raised. For more complex tasks, however, the reverse often occurs and the anxiety level should be reduced. At this level the teacher should present large concepts in smaller chunks. In addition, the teacher should provide activities that suggest status, enjoyment, advantage, and continued use.

Motivation, focusing, and anxiety level become the basic learning principles at the exposure level of the taxonomy. All are extrinsic and require a teacher role dominance in the teaching–learning act. The first (motivation) calls for a wide spectrum of stimuli to pique the senses, the second (focusing) covers the selection and organization of stimuli into a manageable direction, and the third (anxiety level) leads to positive learner response and an observable learner readiness for further progress through the experience. All demand careful and specific teacher planning, for the learner who fails to be motivated, cannot focus on the particulars of an experience, or is under- or overanxious may reject the whole sequence. When the response is positive, however, and the readiness established, the learner is ready for the next level of the taxonomic experience.

## PARTICIPATION

### INITIAL GUIDANCE

In order to reinforce a positive association with the material presented at the outset of an experience, the principle of initial guidance must be brought into play. This helps to keep the learner's anxiety level within bounds and to strengthen the conscious or unconscious decision to proceed. Here the learner needs to know the "rules of the game," the structure within which to operate in order to move ahead. The specific learnings to be acquired or obstacles to be overcome need to be clarified more individually, as do the rewards and advantages of achievement. Thus, by definition, it involves on an individual basis the directions, demonstrations, and monitoring necessary to ensure learner progress and confidence that learning can be achieved. Here the teacher controls the association between the stimulus and the response, the communication employed to direct learner attention, the sequencing of activities, and the feedback the learner receives. Maximum guidance is required at the initial stages of new learning, followed by a gradual teacher withdrawal. Techniques most often used by teachers are walk-through and/or imitation activities; immediate supportive feedback of each step; working with small groups before allowing learners to work independently; and providing for examples, demonstrations, illustrations, or discussion of "what it will look like when the learner gets there." Attention to the principle enables the learner, who knows what to expect, to perform better, to increase the overall speed and the degree of learning, and to continue longer.

### MEANING—EXPLORATION

Although there may be some earlier awakening of perceptions, the teacher should undertake to clarify and to facilitate understanding in the learner at this and at subsequent levels. The learner begins to know the parameters of the experience, to see its meaning in context, and to acquire an understanding of its personal importance. This is an exploratory stage because the learner recognizes that the experience does have substance in that it relates somewhat to past experience; yet it is the touchstone that links the past with the present and that at later taxonomic levels becomes the projection into the future. Meaning–exploration must be ongoing before identity with an experience can come about. There must be a linking of one's personal past with the new meaning before continued learning can take place. Attention to this principle will ensure a regular increase in the rapidity of learning and in the potential for retention, as the learner gains additional experience. The teacher, when possible, should give meaning to the task as soon as possible. This is generally achieved by

- determining the level of understanding in the learner's association with familiar material
- connecting presented material to the familiar by a logical, planned sequence
- clarifying the task set: making certain the learner knows what is to be done and what the completed task is like

The functions of meaning–exploration are to present a preview of success, to increase the speed of learning, and to intensify the degree of retention. It has social relevance as the learner deals with the stimuli and data in relation to others, and it is enhanced when students learn in context with peers.

### CHANCE FOR SUCCESS

With meaning–exploration the learner considers the chance for success. The learner, as exploration and meaning occur, comes to grips with the possibilities for success with the idea, with the activity, and with the experience, assessing personal ability to complete satisfactorily a learning activity. Psychologically, this is also an assessment of the chances for transfer to other contexts. In every learning situation, this is a time to reflect and to project. With meaning–exploration, the past and present have become linked. Here the prospects for the future are considered. The learner must deal with this experiential principle of learning before moving to emotional and intellectual commitments. Again, the teacher helps the learner to see the experience from past to present and into the future. The process involves reason and challenge and exemplifies the teacher's role as a catalyst at the participation category of the taxonomy. A learner must have the opportunity to explore roles, to project ideas and to attempt transfer. A learner who has had these opportunities and accepts them, is ready for activities at the next level of the taxonomy. If the learner does not see a chance for success, the experience terminates until the wise teacher recognizes this and finds a way to resolve it. This is much more difficult on the second occasion. Some characteristics of the principle can be summarized as follows: (a) success enhances those intellectual activities associated with planned learning and failure depresses them; (b) success leads to a raising of the level of aspiration and failure to a lowering; (c) the level of aspiration tends to follow the level of performance but is more susceptible to change by success than failure; and (d) the stronger the chance for success, the greater the probability of a rise in the level of aspiration. Likewise, the smaller the chance of success, the lower the level of aspiration. Teacher techniques involve having students set realistic work goals requiring reasonable effort; assisting the learner through a task or subtask; breaking tasks into component parts, selecting one the learner can achieve and reinforcing the achievements; avoiding both comparisons with the "bests" and remote goals; reviewing past student performance and setting levels of achievement just slightly beyond the level attained; and using the small-step method to ensure successful progress for "defeated" learners. The functions of chance for success are to bring about completion of a learning experience and to increase the level of aspiration.

## IDENTIFICATION

### PERSONAL INTERACTION

This learning principle is one that deals with the learner's movement toward identification with the experience. The learner, through initial guidance, exploration of meaning, and an assessment of the chances for success, has begun an emotional and personal intellectual identification with the experience. This principle involves the learner in active participation with the various components of the experience.

This active participation involves overt or covert responses that increase insight into a learning or problem solution (see Table 5.1). Personal interaction is essential to learning. It means various kinds of field activities, discussion, conferencing, and use of data. Personal interaction can also involve hypothesizing and testing of hypotheses. The learner can become involved in the higher levels of cognitive thinking. The characteristics of this principle are

- interaction of the individual learner with group members, resulting in a pooling of individual resources and thus a greater understanding of total experiences
- increase efficiency of the learning process through learner identification with the experience

Through well-defined teaching strategies (see Chapter 8), the teacher can achieve through this principle appropriate taxonomic goals. Teacher techniques in the achievement of those goals involve both the covert and overt responses mentioned previously. Although covert response is more efficient in saving learning time, the teacher still needs an overt method of telling whether or not the learner is achieving the planned goals and objectives. Too often only the paper-and-pencil method is used. Some examples of teacher techniques are listed. Through these and other techniques enhanced by teaching strategies at the identification level, the student will have more opportunities for permanent change in learning. Personal interaction, furthermore, focuses greater attention on the learning task. It provides a comfortable framework for learner identification with the experience, thus preparing the learner for internalization of the planned experience objectives.

**Table 5.1**
Covert and Overt Personal Interaction Responses

| Covert | Overt |
| --- | --- |
| Assessing | Making choices, tasting, touching |
| Associating | Discussing, signaling |
| Estimating | Signaling, making choices, timeline building |
| Observing | Imitation, reproducing in visual form |
| Evaluating | Selecting, debating |

## KNOWLEDGE OF RESULTS

This is another learning principle fundamental at the identification level. It provides the learner with an insight into the achieved degrees of success or failure. Here the learner becomes aware that the experience has changed personal behavior and knows what has happened. Outcomes are anticipated at this level. The results, rather sketchily outlined in sequencing, now become apparent and involve learner recognition of the process. The whole experience can now come into focus. The teacher's role is to assist the learner in seeing the results and then to sustain the modified or changed behavior. Great learner satisfaction and/or learner pride can be seen as knowledge, and recognition of results become apparent. Knowledge of results begins at the exposure category, but, like pieces in a puzzle, begins to fit together and finally to take shape at this taxonomic category. Knowledge of results really provides the learner with an insight into the degree of success of failure. It has the following characteristics:

- The closer to the completion of the learning task or subtask, the more effective it becomes.
- Feedback must be specific to improve the performance or learning.
- It establishes connections that differentiate between learning situations.
- It limits effect of other stimuli.
- It should take place before practice or overlearning can be successful.

Some techniques for helping students achieve knowledge of results are immediate correction of work, having students compare their work with a correct model, signaling approval, and setting standards that the student understands prior to the performance of the activity. Knowledge of results serves chiefly to increase precision of learning and to increase the possibility of the development of intrinsic reward.

## REINFORCEMENT

This is a transference within the learner from extrinsic motivation, dependent on extrinsic stimuli and teacher planning and manipulation, to intrinsic, self-progressing learning. It is a condition that follows a response and that results in an increase in the strength of that response. The experience in participation activities brought the past of the individual together with the new stimuli and, as the experience was personalized, the learner could decide to identify with the experience. This is a very subtle process and one that involves real learner effort and teacher skill. It is only with identification that the learner can move toward internalized behavior. Reinforcement becomes a major and fundamental learning principle once this identification is made. Reinforcement involves choice to continue to learn and teacher ability to change roles from catalyst to moderator. In effect, the teacher's role is no longer that of the extrinsic manipulator but rather that of the moderator of an accepted experience. When the learner makes an emotional and/or intellectual identification, the teacher must have some plan for the initial and subsequent reinforcements. Conditions that reinforce vary with the learner and are best maintained through the use of a schedule of reinforcements. It should be noted that the successful completion of a task or the discovery of a solution may be a reinforcer for behaviors leading up to these events. Additionally, when reinforcers satisfy felt learner needs, more effective learning takes place. It should be remembered that the magnitude of reinforcement is of less significance than the fact that reinforcement will or will not occur. Steps a teacher can use to augment reinforcement are to (a) identify the behavior to be reinforced; (b) decide what constitutes positive or negative reinforcement; (c) devise strategies to develop and reinforce the new behavior and learning; (d) decide whether to use negative reinforcement to extinguish or suppress previous learning; (e) develop strategies to get the learner to practice new learning and behavior and to reinforce the behavior positively on a regular schedule; (f) remove negative reinforcement, after practice with positive reinforcement; and (g) change to an intermittent schedule of reinforcement with increasingly longer intervals. The functions of reinforcement are to ensure, to remove, and/or to internalize learning. Reinforcement occurs at the identification level and continues to be important through internalization.

## INTERNALIZATION

### OVERLEARNING

At the internalization level, overlearning becomes an important aspect of the learning process. The learner has evidenced some mastery of skills and/or attitudes, and overlearning provides additional scope and influence when something has once been learned. The learner becomes thoroughly familiar with the components of the experience. He has interacted with them, has attributed meaning to them, has projected future implications, and has sequenced a series of projected activities. Through all of these various activities involving many teaching strategies and learning principles, the learner practices, uses, and repeats elements of the experience until they become a natural part of individual activities. Skill learning is strong at this level. When this happens, the learner begins to internalize the elements, and the planned modification of behavior is at hand. This process begins at identification and continues through internalization. It usually involves literal or figurative sharing and interaction with peers and/or teachers. It should be remembered that this process involves periodic reviews that are necessary for permanent retention. These reviews should be at increasing intervals of time. A teacher may have students practice a skill in two or three different activities after it is acquired. Subsequent lessons should provide for the practice of a skill or the application of a concept at increasing intervals. The function of overlearning is reinforcement and retention as well as expansion and fusion of the experience into the life-style of the individual. It eliminates or minimizes the chance that the early learning was a guess, a coincidence, or not "fixed." Overlearning, furthermore, increases the chance for transfer of learning to other situations.

*INTRINSIC TRANSFER*

Whenever the existence of a previously established habit (skill or concept) influences the acquisition, the performance, or the relearning of a second habit (skill, process, or concept), intrinsic transfer takes place. Intrinsic transfer may be positive or negative. When it is positive, the acquisition of one "learning" facilitates the acquisition of another. When it is negative, the acquisition of another is deterred. At this point in the learning experience, the new behavior appears in different contexts. At the point of intrinsic transfer the learner personally transfers the newly learned behavior to other aspects of his activities. This is a personal broadening of the experience in order to see how new behaviors work in a variety of contexts. This principle cannot come into play until the learner has reached the internalization level of an experience. The learner can then recognize this transfer. The learner sees relationships in a variety of ways and acts upon them. The teacher can expedite the process by helping the learner to see the behavior in progressively broader contexts through a series of analyzing, summarizing, and generalizing activities.

## DIFFERENTIATED INPUT

Differentiation involves changing the quality of stimuli to further the movement toward an incorporation or an unconscious level of behavior. It activates the learner to control his or her own progress by changing the dimensions of the experience (appearance or environment) and the characteristics of the stimuli (words, pictures, or activities). The discerning teacher will now change the level of stimulation with the intent of both expanding the scope and ensuring the fusion of the experience. Familiarity with the learners and their developmental and maturation levels will assist the teacher in selecting activities for differentiation of the experience. Knowledge of the characteristics of differentiated input will make this selection easier. These characteristics are summed up as follows: Human sensitivity changes as it becomes accustomed to stimuli (routine); any novel, different, or unusual occurrence causes a re-alerting reaction; the magnitude of change is not related to the degree of learning. Teacher techniques for implementing differentiated input include checking the learning task to see what change in the teacher or learning strategy is necessary to maintain interest or attention, and making changes in the characteristics and/or dimensions used previously. As stated, the change need not be extreme in order to further student interest. For example, simply changing schedules, using games, pairing students to work together, or any slight break in the routine may be enough for younger students, whereas placing the experience in a problem-solving, creative, comparative–contrastive, or novel situation may be necessary to maintain the focus for older ones.

## DISSEMINATION

## EXTRINSIC TRANSFER

Extrinsic transfer in the experiential taxonomy means communicating to, influencing, or teaching others about the experience. It is the social context of intrinsic transfer and involves learner choice. Extrinsic transfer cannot be imposed; the learner does it voluntarily. In effect, the learning experience is not complete in a social context until this transfer is made. No one "is an island entire unto himself" and changed behavior has little efficacy until it is demonstrated in the social milieu. The teacher here assumes the role of critiquor with the major purpose of keeping the cycle of learning going beyond the existing setting to the exposure level in a new learning context. Learning can be seen as cyclical—one set of learnings opens the door to new learning experiences that begin anew at the exposure level. The teacher, in an evaluative–critical role, has resumed the dominant role of the exposure–participation categories of the experiential taxonomy.

## REWARD

This is the intrinsic aspect of reinforcement and extrinsic transfer. Although often apparent at identification, particularly with the learning principle of knowledge of results, this learning principle is more properly a dissemination aspect. Reward can be intrinsic or can be expressed in the process of dissemination to others. During the dissemination stage and when extrinsic transfer and reward become dominant learning modes, the teacher's role becomes that of a vehicle whereby the student can express the experience and then that of the critiquor and evaluator. In this sense it becomes the teacher's responsibility to ensure that the learning continuum does not stop but instead continues to provide opportunities for the learner to assume the teaching roles and to carry on the continuum. In the role of constructive critic, the teacher, through corrective, supportive, and informational feedback, brings in new stimuli and new experiences so that the learner can build a new continuum based on the present stage in the learning experience and on the newly acquired behavior. In effect, the whole process has been recycled.

## INTRINSIC MOTIVATION

The epigenesis of intrinsic motivation from extrinsic motivation is gradual. An awareness of striving toward satisfaction accompanies most motives. Behaviors, themselves, can often be predicted with a fair degree of accuracy, and this forms the basis for what has been previously termed extrinsic motivation. With more complex advancement through the taxonomy, internal stimuli occur. Opinions, attitudes, beliefs, and, most importantly, values frequently change. There is often an inner pairing of activity with satisfaction, of behaviors with values, of drive with achievement, and of persuasiveness with recognition. Thus, motives can find expression entirely without awareness or, when there is consciousness, without clear basis. Motivation at this level means that the drive, the need, or the desire operates from within and the learner now automatically seeks to manipulate the variables (persons, task, conditions). Unlike those for extrinsic motivation, reinforcers can be intermittent and delayed and are most often intrinsic themselves. Whereas intrinsic motivation neither ensures that the learner will reach the homiletic state nor is a requirement for this, it is not possible to conceive of a person sustaining this level without it. The teacher's role now is one of critiquing the strategies being used and redirecting the learner to new experiences that build upon the old.

**Table 5.2**
Basic Taxonomic Learning Principles

| Taxonomic level | Teaching role model | Learning principles |
|---|---|---|
| 1.0 Exposure | Motivator | 1. Extrinsic motivation<br>2. Focusing<br>3. Anxiety level |
| 2.0 Participation | Catalyst | 1. Initial guidance<br>2. Meaning–exploration<br>3. Chance for success |
| 3.0 Identification | Moderator | 1. Personal interaction<br>2. Knowledge of results<br>3. Reinforcement |
| 4.0 Internalization | Sustainer | 1. Overlearning<br>2. Intrinsic transfer<br>3. Differentiated input |
| 5.0 Dissemination | Critiquor | 1. Extrinsic transfer<br>2. Reward<br>3. Intrinsic motivation |

**Table 5.3**
Learning Principles: A Taxonomic Continuum

| Teaching principles | Exposure | Participation | Identification | Internalization | Dissemination |
|---|---|---|---|---|---|
| Extrinsic motivation | Extrinsic–sensory[a] | Extrinsic, data gathering | Decreasing, research extrinsic | Limited | Limited |
| Focusing | Sensory–stimuli, use of media and/or of variety of stimuli to identify particular elements[a] | Modeling and imitating particular elements | Research and conferencing, continuing use of media | Skill reinforcement, lab activities, use of media | Group dynamics |
| Anxiety level | Pleasure–pain reacting[a] | All strategies, as level of risk increases | All strategies though anxieties diminish | All strategies, some increase due to confronting ideals, beliefs, and values | Variable dependent on feedback |
| Initial guidance | Stimulus–response | Teacher-directed participation in a wide variety of activities | Teacher, student, and peer interaction for developing research sequence | Role playing–simulation; interactive evaluation is the chief manifestation of initial guidance | Presentation–reporting |

| Meaning–exploration | Response from stimuli also interaction–questioning | All strategies, dramatic play an important one to lend meaning | Research and hypothesizing are strongly linked to this learning principle | Any of the strategies particularly summarization and comparative–contrastive analysis | Seminar is a way of showing meaning |
| --- | --- | --- | --- | --- | --- |
| Chance for success | Readiness for next levels of taxonomy | All strategies, particularly dramatic play in which learner projects next step[a] | Sequencing | Knowledge of achievement | Teaching or sharing with others |
| Personal interaction | Stimulus–response | A wide range of teacher activities including modeling, imitation, and all other strategies | Discussion takes place but can be manifested by other strategies as well[a] | Role playing is the chief manifestation of personal interaction here | Presentation–reporting |
| Reinforcement | Same as response to stimuli where there is linkage to other experience | Modeling and imitation as well as other strategies | All strategies here as past, present, and future possibilities come together[a] | Skill reinforcement manifested in many ways | Dramatization and other presentation–reporting situations |
| Knowledge of results | Readiness for next level | Exploration of parameters of experience | Generalizing about result of research[a] | Awareness of how behavior and/or results have developed | Interaction in the presentation–reporting mode |

(continued)

**Table 5.3—*Continued***

| Teaching principles | Exposure | Participation | Identification | Internalization | Dissemination |
|---|---|---|---|---|---|
| Overlearning | Some in continuing stimulus–response | Shown as it relates to past experience | Research, hypothesizing, and all strategies involve overlearning | Skill reinforcement[a] | Displayed in any dissemination strategy |
| Intrinsic transfer | Limited | Seeing experience in terms of own past | Research and hypothesizing | Modified behavior[a] | Any strategies |
| Differentiated input | Readiness for next level | Exploration of parameters of experience | Using many data to begin to generalize | Awareness of how behavior has changed[a] | Interaction in the presentation–reporting mode |
| Extrinsic transfer | Limited | Limited | Sharing with others | Interaction with others, summarization | Influencing, teaching, or reporting to others[a] |
| Reward | Response to stimulus | Discovering parameters | Emotional and intellectual identification | Pride in accomplishments | Sharing pride in accomplishments[a] |
| Intrinsic motivation | Undetermined | Undetermined | Initial intrinsic research, hypothesizing | Intrinsic, achievement of defined behavioral objective | Intrinsic, re-creation[a] |

[a] The fundamental or essential category for each learning principle.

# AN OVERVIEW

Table 5.2 is an overview of learning principles fundamental to each category of the experiential taxonomy. It shows those learning principles that are necessary and central at each particular category. The selected learning principles do, of course, sustain themselves throughout the taxonomy, manifesting themselves differently at each taxonomic level. It should be noted that for each taxonomic level, we have designated a teaching role model. These role models are explained in detail in Chapter 6 on role models.

## LEARNING PRINCIPLES: A CHART OF TAXONOMIC SEQUENCE

As noted earlier, learning principles have an ongoing, unique quality about them. Though they may be related and utilize the same or similar teaching strategies, they have little or no overlap. Table 5.3 follows each teaching principle through its taxonomic sequence. This, in succinct form, shows the development and consummation of a learning principle from exposure to dissemination. One will note that most learning principles have some particular manifestation at each taxonomic level. Examples of this are intrinsic transfer and extrinsic transfer, which are relatively unimportant at the early levels of the taxonomy. Each learning principle may be effective and useful at most or all taxonomic levels though it is fundamental at particular ones. For example, motivation is fundamental at exposure but is used at all taxonomic levels. One should also note that the most effective learning sequences will combine several or even all of the learning principles, in a wide spectrum of teaching strategies in the classroom.

# 6 | A Taxonomic Role Model for Teachers

One of the major questions in education in recent years has been that of the teacher's role in the classroom. Various models and role definitions for each have been espoused or presented and many of those have proved viable in the teaching–learning act. Yet much remains to be learned about teaching roles. In a recent request for research proposals, the National Institute of Education noted that "the history of research on teaching, together with the current emphasis on competency-based certification, implies the need for renewed and revised efforts to define effective teaching [Basic Skills Research Grant Announcement, Spring, 1976, p. 3]." With this in mind, we offer this taxonomic role model for teachers as a new approach to the definition of effective teaching.

The experiential taxonomy as defined in this work is a simple yet straightforward, categoric formalization of human experience. With its gestalt view of human experience, it provides a unique vehicle for teachers in their work. An examination of five major categories of the taxonomy can lead to a clarification and understanding of changing teacher roles throughout the teaching–learning act. Teaching roles change as the learner progresses through the teaching–learning process.

Role models for teacher's have not, for the most part, been taxonomically sequenced in the past. Most role models explicated one dimension of the teacher role and adopted that role definition to a variety of situations. Unfortunately, that type of one-dimensional role will not always function because teachers change roles as experiences change and as learners move progressively through teaching experiences. Then, too, in a group situation learners may be at different experience levels, so the teacher must be flexible in dealing with specific subpopulations of the group. In this experiential model it is our position that functions do change and that at each taxonomic category of an experience there is a recognizable shift in the emphasis of the teaching role. As learners progress through the activities involved in an experience, these role changes can be defined. Briefly, they are as follows:

| | |
|---|---|
| EXPOSURE: | Motivator |
| PARTICIPATION: | Catalyst |
| IDENTIFICATION: | Moderator |
| INTERNALIZATION: | Sustainer |
| DISSEMINATION: | Critiquor |

For purposes of this model we will use only the major categories of the taxonomy. We will not define roles in terms of subcategories of the five levels.

Each of these taxonomic roles represents a different approach to and facet of the teaching–learning act. A teacher who is aware of these changing functions, can plan activities and interaction according to the needs of the learners. The teacher can choose from those teaching or interaction strategies appropriate to the teaching roles at each taxonomic level. A step-by-step curriculum development plan and a listing of taxonomically keyed teaching strategies are noted in Chapters 3 and 4. In this chapter, the purpose is to present a brief definition of the teaching role at each taxonomic level in order to enhance total teacher preparation and planning and so that the changing teacher role can be better understood and executed.

## EXPOSURE

The major thrust of the teacher role at this level of the taxonomy is motivation. The teacher as extrinsic motivator is essential to the teaching–learning act at this level. It is the task of the teacher, first, to establish communication with the learner so that motivation can begin. At the same time, the teacher must extend the invitation to learn or to experience in such a way that the learner accepts it. Thus, not only must the goals and objectives at this stage be planned and kept well in mind, but the activities must also be carefully selected. Furthermore, the expected ultimate outcomes in terms of modified behaviors must also be carefully planned with the specific outcomes noted and listed.

In the teaching situation, the environment should be arranged to provide a stimulus for beginning an experience. The invitational role of the teacher is to provide stimuli through a variety of activities and through the arranged environment. From those activities there should be an expected, detectable learner response. In the development of a multicultural unit, for example, the teacher's ultimate purpose may be to have the students demonstrate an understanding that several selected cultures are similar in one or more of the aspects of culture such as trade, architecture, literature, and government. In planning the exposure or imitational stage one activity is chosen in which the teacher shows pictures of the selected cultures in order to define terms that the students will encounter. To provide the necessary stimulus, however, the teacher assumes the role of a "teller of little stories" while showing the pictures. Thus, the presentation is not so much a lecture as it is a little biography of what is to come. In addition, it achieved one of the teacher's objectives at this stage, which was to create an awareness of terms and shades of meanings of which the students were unaware.

Wherever possible the teacher should examine learner strengths, interests, and limitations in order to choose the activities more appropriately. The invitational role is paramount in the learning setting. The function is directive and requires that the teacher actively manipulate the environment. In this initial taxonomic setting, the teacher's role is more dominant than at any succeeding taxonomic level. The teaching task is to "make" the learner want to move to other categories of experience. The imperative "make" does not imply that the teacher need be threatening, arbitrary, or demanding. Indeed, part of the teacher's role as a motivator is to establish a positive and healthy rapport with the learner and to reduce any anxieties that may be anticipated during planning. There are, of course, many motivational techniques and strategies designed to move the learner into deepening experiences. Such teaching techniques as demonstration, use of media, directed observation, promoting problem identification, and interaction–questioning are appropriate for motivation. It is essential at this taxonomic level for those techniques or strategies to succeed; for if the motivation fails, the learner will probably not achieve the expected or modified behaviors in this experience. All the skills a teacher possesses should go into this motivational role. If one strategy does not work, it is incumbent on the teacher to have a backup plan in reserve.

The teacher should very carefully and specifically plan all strategies and techniques of motivation so that the learner can move from an initial response to the motivation toward a demonstrated readiness for participation. No other category or role model in the teaching–learning act requires the specific and careful planning that this one does. When motivation has been achieved, another dimension of the teaching–learning act becomes apparent.

## PARTICIPATION

Once the learner has demonstrated readiness for participation in the experience, thus moving to a new taxonomic experience level, the teacher role changes and the teacher assumes a new role, that of a catalyst. By definition, a catalyst is an agent of change, something or someone that speeds up the process of change toward planned objectives. At this point, a teacher brings a spectrum of resources that challenge the learner and form the bridge between an initial awareness and actual identification with the phenomenon. Here the teacher engages in dialogue or provides activities that introduce to the learner the information and data essential to the experience. Parameters of the experience are explored through the dynamic interaction between the learner and the resources brought by the teacher. A variety of questioning techniques and information-giving strategies are appropriate at this level. The teacher, in this role, must be both a repository for and a dispenser of data. The task must be to sharpen the perception of the learner so that there is a readiness for an emotional and intellectual commitment to a deepening and ongoing experience.

As stated in an earlier chapter, the learner's acceptances of further exploration into the experience is now of greater risk than before since it now involves changes both in learner behavior and in learner perception. Thus, the teacher must "signpost" and carefully guide the learner in order to perform successfully the catalytic role. The learner needs to know what is to be done, how it is to be done, and what the product should look like when it is completed. Referring back to the multicultural unit mentioned at the exposure level, it may be that the teacher at the identification level wishes the students to develop their own retrieval chart illustrating the similarities and the differences among selected cultures. To accomplish this, however, the teacher must first carefully walk the students through the steps using the data acquired during the invitational stage. This not only forms the bridge to the next level by promoting both recall and simple data-gathering activities but also increases the chance for success in achieving known goals and reduces constraints to the learning process. Thus, it is pertinent that the learner now become aware of some of the expected outcomes and establish a pragmatic need to continue the experience.

A wide variety of activities including brainstorming, modeling, sequencing, dramatization, and dramatic play are appropriate for the student at this level. These and other strategies will involve the learner in the experience. The teacher is the catalyst who brings the parameters of the activity to the past experience of the learner and is the agent by which the learner modifies the experience and begins to identify with it. The teacher role is still that of an extrinsic agent. The experience has yet to become intrinsic to the learner, and the teacher's function remains somewhat manipulative in terms of the learning situation. The teacher is more active and manipulative in this role whereas the learner tends to be somewhat more passive, though in learning the possible outcomes and discovering more about the experience, the learner begins to take a different perspective regarding it. The experience is still outside the learner's intrinsic motivative behavior. This, however, begins to change as the learner starts the process of identification with the experience.

## IDENTIFICATION

Identification is the taxonomic level at which the learner makes an emotional and intellectual commitment to the experience. It becomes "my experience," and the learner is ready to experiment with a variety of possible behavioral responses to this situation. This important, and sometimes very subtle, emotional and intellectual commitment of the learner moves the experience from an extrinsic one to an intrinsic one. This is the stabilizing stage. The learner with great guidance has moved through the experience. The teacher tries to enable the student to do the task or to perform the skill with less help. The learner is now aware of how, of where, and of when—in one or two instances—the learning and the experience are useful.

Thus, at the identification level, the teacher appears somewhat less active and becomes more passive. The teacher's role is, nonetheless, vital and essential. The responsibility for experience begins to shift at this point to the learner. The teacher, in effect, becomes a moderator and must keep things running smoothly and coherently. Whether the teacher's style is traditional, open, or otherwise, the role is to make certain all the collected data tie together. In addition, a good moderator frequently must be creative and inventive to maintain the established momentum. Furthermore, the teacher must not only know the experiential background of each student but now must also record the progress being made.

At the identification level, the teacher must moderate questions, provide options to explore, arbitrate decisions, suggest different modes of operation, and basically keep the learner focused on the experience and moving toward the planned internalized behavior. When these things occur it becomes apparent that the teacher's role has changed. From a dynamic motivator to an active catalyst, the teacher becomes a subtle but inventive moderator whose job it is to bring reinforcement, personal interaction, and knowledge of results to the learner. This does not mean, however, that the teacher's presence and role do not have an impact on the learner. The goals are to clarify issues, to keep the learner within the parameters of the experience, and to suggest paths and ways of achieving the planned internalized behavior.

This role change may sometimes be difficult for many to assume because of some rather traditional concepts of teacher roles. It really serves to point out the need for great flexibility on the part of the teacher. The teacher must sense the change in role from dominant and manipulative to more subtle and moderating. This requires a sensitive and insightful person capable of relinquishing more obvious manipulation and control yet able to assume a keen capability for moving the learner to complete the experience. When this level in the teaching–learning act is achieved, the learner must not feel alone in the experience. A great deal of peer–teacher interaction and peer–peer interaction is appropriate at the identification level of experience in order that the learner may see the activities in a social context as well as from an individual and personal perspective. At this point group conferencing is very effective as is committee work and discussion in the classroom. This sharing for identification and commitment leads to the next level of experience and to another role shift for the teacher. The learner, upon making an emotional and intellectual commitment to the experience and upon being involved in peer sharing, begins to gain some knowledge of the expected results and begins to internalize the behaviors that the teacher has so carefully planned at the exposure level of the taxonomy.

## INTERNALIZATION

In this taxonomic category, the learner starts to demonstrate the preplanned behavior. Through differentiated input the student begins to act on the data presented and on prior interaction with the experience. The teacher should not be disappointed if there are learner interpretations of internalized behavior differing slightly from the planned behavioral outcomes. Learners have their own past experiences to bring to the experience. In fact, this should be positively accepted when appropriate. When new behaviors begin to occur, however, the role of the teacher shifts again to that of a sustainer who is supportive of the newly modified behavior. The modified activity patterns exhibited by the learner are the results of the individual learner's behavioral response to the motivational environment, participation in the experience, and identification with it. It is the task of the teacher to see that this behavior becomes more than just a particularized act having little permanence in the learner's life, that these demonstrated behaviors become instead more permanent, even a lasting "modus operandi" for the learner. The "sustainer" should help the learner recognize the behavior as positive, appropriate, or necessary and then should provide rewards for the learner in a supportive and upholding role. The teacher, although more passive than at previous levels, must now move to extend the learner's experience both inwardly and outwardly with the learner in more control than before. At this level of experience activities providing additional scope and familiarization in a variety of situations can increase the chance not only for permanency but also for transferring the learning to other areas of learner behavior. The teacher now structures the environment for expanded skill practice, laboratory experiments, role playing, simulation, and originating and summarizing activities. These can, and in most cases should, result in overlearning so that the learner can begin to see the results of a personal relationship with the experience. Behavior that is not sustained by peers, by society, or by a respected authority seldom becomes permanent in a human life. It should be noted that at the very core, the teaching act is basically value-oriented. It is the unique role of the teacher to support and to encourage specific planned behaviors and to create a sustaining environment so that the learner can make the internalization of the behaviors long lasting and even permanent.

This teaching role is rewarding in that the teacher can see the achievement of carefully planned goals and objectives. When this experience level is reached, the mutual support and rapport between teacher and learner can really be sensed. It is the trust and the mutual respect that so enhance the teaching–learning act. Successful accomplishment of this step in the continued modified behavior of the learner is a vindication of careful planning and of the teaching–learning act. There is, however, yet another taxonomic level and another role for the teacher to assume. The role comes at the final level of experience, that of dissemination.

## DISSEMINATION

Internalized behavior is disseminated when the learner through some outward expression signals overtly the degree of transfer, of reward, and of motivation achieved. Dissemination differs from the sharing aspect of the identification category in that at this point the learner recognizes the modified behavior as having an impact on his life and desires to let others know. This expression leads to a new beginning of the experience cycle, a continuum that never ends. Furthermore, when carefully examined, it suggests to learner and teacher alike the way to new learnings. The teacher has moved to the role of critiquor and, once the expression is made, has the task, indeed the obligation, of renewing experience and of helping the learner put the experience in a new context. As critiquor, the teacher evaluates the learner's readiness for dissemination, juxtaposes new data, and attempts to expose a new series of stimuli to the learner. The new behavior must be tested in terms of these new stimuli and new circumstances. In effect, the experience has made a full circle and exposure begins anew with the learner as motivator. One should note also how the role of the teacher moves from dominant to more passive and then, at the end of the dissemination level of the taxonomy, again becomes dominant. Initially, the responsibility for moving the learner from one behavioral pattern to planned modified behavior was totally the responsibility of the teacher. While moving into deepening levels of an experience, the learner must assume more responsibility and must take a more active role in progressing toward modified behavior. At the completion of the dissemination level, however, the teacher role again becomes dominant as the shift to new kinds of experience occurs. The awareness of the need to shift in role responsibilities requires a mature and insightful teacher who should very sensitively grow into this role as a result of continuing interaction with the learner.

SUMMARY

These, then, are the taxonomic roles of teachers. They represent the sequence of pedagogical roles intrinsic to the total teaching–learning act. Of course, they are not constant. In a class or even in individual learning sessions, the roles vary and fluctuate according to the unique and individual patterns of human interaction. With careful planning and sensitive reaction to the vital dynamics of teaching, this taxonomic role model can prove to be an effective and responsible tool for teachers and counselors committed to helping learners become more effective in their ongoing societal roles.

# 7 | The Taxonomy and Creativity, Critical Thinking, and Problem Solving

It was noted previously in the chapter on designing curriculum that, whereas creativity, critical thinking, and problem solving are generally included, curriculum writers desiring to maximize the importance of these areas should key them to the different levels of the taxonomy. Performing this additional step for whichever area it is deemed essential frequently causes the planner to revise or to modify the activities originally selected and, thereby, to strengthen the units. In addition, inclusion of all three areas prompts the planner to provide at each taxonomic level additional activities that not only meet the different learning styles but also provide for the various federal and state requirements demanding provisions for "over and above" activities for specific students. In brief, the taxonomic sequences for each of the three areas are as follows:

## Creativity

EXPOSURE:              motive to produce
PARTICIPATION:      visualization of production
IDENTIFICATION:     experimenting with the idea
INTERNALIZATION: completing a product
DISSEMINATION:     admiring, showing, sharing

## Critical Thinking

EXPOSURE:              recognizing variable
PARTICIPATION:      data collection and variable definition
IDENTIFICATION:     organizing, structuring
INTERNALIZATION: generalizing
DISSEMINATION:     using variable

## Problem Solving

EXPOSURE:              problem identification
PARTICIPATION:      exploring data bases
IDENTIFICATION:     trying optional solutions
INTERNALIZATION: selecting a particular solution
DISSEMINATION:     implementing and influencing

The sequences for the three areas, although defined separately in this chapter, are closely related even though the curriculum planner may wish to focus only on one. For example, as one progresses through the creative sequence, there is concomitant progression through critical thinking and through problem solving. Each has, however, an integrity that is identifiable and sequential, and each has been treated separately in this presentation. This separate presentation serves not only to define the unique processes involved in creativity, in critical thinking, and in problem solving but also to help curriculum planners, administrators, and practitioners deal with each in the practical preparation of successful learning experiences. In this way, learning can be fostered and the progressive steps through the taxonomy can be recognized, encouraged, and brought to fruition as part of the total experience package.

# CREATIVITY

Creativity is a personal interaction with an idea, with material, or with a problem. It is a process that requires sequences and activities unique to the individual and that results in a product, an acquired skill, or a modified behavior. As much emphasis should be placed, however, on the process of creativity as on the product. It is in the process of creativity that an individual begins to enhance self-image and to feel a sense of accomplishment. The product is only the culmination of the process. It is a practice in which all of us participate and that all of us recognize. It is natural and is a part of our daily experience. We concur with Gordon (1961) in his assumption that the creative process is not mysterious. Whatever is new to the individual is creative for that person. Unfortunately, society has narrowed the definition of creativity by putting a premium on certain products of creativity and has ignored the natural creativity all people possess. This, thus, creates an elitism for the successful practitioners of socially approved creativity and makes most people feel they do not possess this trait.

Creativity, as a personal interaction with an idea, with material, or with a problem, is an initial, productive activity that results in some artifact or object, in recognized behavior modification, or in some idea that others may recognize. Creativity, therefore, is not necessarily something completely new in terms of human experience but is rather something that is unique to each individual. As personal horizons widen, as one becomes involved with new ideas, new occupations, new social relationships, or new environments, the creative process is involved. In this sense, the most creative years of one's life are those early years when the young child is experiencing rapidly expanding environmental and human interaction. Creative experience, however, can be fostered in the school and does persist through life.

Viewed as a sequential process, creativity can be seen as a series of progressive steps that one may note, foster, and plan. Its activities can be identified and ordered and, therefore, can have a taxonomic sequence. We posit that a sensitive teacher using a carefully planned sequence can prepare for creativity, nurture it in students, and build a classroom environment where creativity happens and is recognized.

Creativity is above all a gestalt experience and the experiential taxonomy is an appropriate vehicle for developing its sequence. It provides a framework for deepening levels of creativity in a logical and recognizable way. Through this taxonomic sequence, curriculum planners can define, sequentially, the creative process and can plan better for its classroom use. In using the experiential taxonomy for sequencing this area, only the major categories will be utilized definitionally.

*EXPOSURE*

At this initial level, the teacher provides a climate for creativity through motivation. In the classroom environment the teacher builds an atmosphere in which the learner feels a need to produce something, to express something, to learn something, or to do something. An arranged environment can provide sensory stimuli, and learner response to those stimuli can produce within the learner a felt need to continue with the experience and to react in some productive and creative way.

At this point, the conditions of creativity are important. In terms of the classroom, the arranged environment, the various media, and the type of teacher–student interaction are significant. Deprivations or surpluses can expose the individual to the need for creativity. Problem identification, comparisons, and the recognition of a change from the normal also foster the need for creativity. At the exposure level the learner should recognize that there is something more, something beyond stimuli, beyond an initial need, or beyond the apparent problem. The learner needs to wonder about something. When this sense of wonder is motivated, the learner feels impelled to act and is ready for the next level of creative activity. The role of the teacher is that of a motivator who arranges stimuli, plans for varied responses, and interacts as necessary to produce the need to create. The teacher is the manipulator of the environment and, at this point, is dominant in the experience. The student is the attender, and the most prominent evaluation mode is that of observing learner readiness for movement to the next taxonomic level.

*PARTICIPATION*

At this level of the taxonomic sequence, the learner begins to visualize possible outcomes of the creative process. There is a mental reproduction or representation of the kind of product, artifact, solution, idea, or behavior sought. The learner says, in effect, "I am going to paint a picture" or "I will do this myself." There is a recognition of the parameters of the creative milieu. The task and the sequence of achievement begin to be known. The learner at this level of creativity begins to deal personally with the motivational stimuli. One may visualize, dream, role play, practice, or rehearse the felt creative need. At the participation stage, past experience is used to modify the stimuli, to clarify the parameters of the creative product, and to recognize the process for achievement. This is a time to explore, to discover, and to make a decision to create a particular item. It is here that the participant decides to go on with the creative act. When the learner is at this step in the creative process, the teacher needs to provide resources for exploration and for interaction. The learner, often reacting to the stimuli of the exposure level, now needs material and data with which to interact as a basis for creating. The role of the teacher is that of a catalyst; the learner is an explorer. Personal interaction between teacher and learner is as crucial at this level as is interaction between the learner and the various media. The role of the teacher as a catalyst bringing the learner together with the environment and the resources develops a new set of conditions, and the learner progresses another step in the process.

*IDENTIFICATION*

At this taxonomic level, the learner, in experimenting with the idea, product, artifact, or behavior, expresses creativity. The learner may put an idea down on paper and edit it, or may use one particular form of expression and then another. The participant needs time to ruminate, to think through alone, and to experiment. This rumination is both active and passive. It involves thinking and doing. Learners need the opportunities for rumination during this stage of the creative process. Knowing the parameters of the creative milieu, the learner now begins to select options, to explore possibilities, and to deal with probable solutions. This identification with the creative process brings about a reinforcement of the need for creative expression as well as an emotional and intellectual commitment to resolving this need. With this commitment, the learner becomes self-motivated to continue the creative process. It is at this level that the learner also begins to share the creative process with others. The commitment has been made and the learner is able to communicate informatively. This sharing can reinforce the creative process or can amend it in terms of additional options or possible directions. Implications for teaching are readily apparent. The teacher continues to provide enriched resources for the learner to act as a moderator between those resources and the learner, whereas the learner assumes the role of experimenter. Furthermore, the teacher must allow for free and individual expression of identification. Once the emotional and intellectual commitments have been made by the learner, the teaching role changes. As a moderator, the teacher becomes a supportive observer, questioning to widen parameters and providing the feedback and resources as necessary.

Sometimes the role of moderator is not easy for a teacher to assume. At the early stages of creativity, the teacher is dominant in motivating, encouraging, and suggesting and is the catalytic agent in the creative process. At the identification level, the teacher and student roles reach a balance; the student becomes more active whereas the teacher works only to help the learner clarify ideas, processes, and data. The teacher's role is perhaps even more vital at this stage because the learner is given more responsibility, even the right to make mistakes. Creativity sometimes stops here because teachers cannot adjust to this new role of moderator, which creates a more permissive atmosphere in the classroom area at this stage than at any other level of the creative process. In the identification stage, the learner begins to narrow the options and to select the most logical and effective solutions. As this winnowing and narrowing process emerges and continues, another level of the creative process becomes apparent, and the product of creativity begins to take shape. The learner is now ready to move into the next taxonomic step in creativity when the product will actually come into being.

### INTERNALIZATION

This is the step in the creative process when the learner, after narrowing options and choosing possible solutions, completes the creative process and achieves a product. The product is finished, the idea formulated, the action done, the picture painted, or the composition completed. Something new to the experience of the learner has been accomplished. Though this product may be amended, edited, or changed, a new experience has occurred, and it can be defined as a unique and creative accomplishment. The learner has internalized the process of creativity and experiences the intrinsic satisfaction of doing something new for the first time. The learner manifests modified attitudes, skills, or behavior as a result of this creative process. An enhanced self-esteem is often the result of recognized achievement. The learner needs to be aware that someone else is pleased with the result of his creativity. The role of the teacher during this step of creativity is to sustain and support manifested behavior modification as the learner demonstrates the internalized creativity. The teacher, thus, becomes the warm and supportive approver and sustainer of the changed behavior. Appropriate praise or reward should accompany the creative act, the product, the new idea, and the observed behavior changes. The teacher, as sustainer, functions in a positive sense to encourage learners toward dissemination when that is appropriate.

## DISSEMINATION

Even though the learner has experienced intrinsic awareness of a completed task and there is an observed behavior modification, the creative process is not completed in an extrinsic context until there is a dissemination of the creative act. On the part of the learner, dissemination involves an attempt to be the presenter, the demonstrator, or the motivator so that others will appreciate the product and/or have a similar creative experience. Furthermore, the learner may teach someone or may homiletically disseminate an idea. This is the extrinsic dimension of the intrinsic creative product or new behavior. The point of this role is to renew the creative process by beginning again in the creative mode. In a true sense, the teaching–learning act has come full circle. Classrooms and learners thrive on creative experiences. Teachers, likewise, grow as they become both vicariously and actually a part of the excitement of creativity. This experience is a healthy and needed expression of growing human activity, for it is in this process that the individual grows and becomes more effective in social interaction. Creativity, in this sense, is not esoteric and eccentric but, rather, is a vital and universal human function. To use Whitehead's term, "It is the universal of universals characterizing ultimate matter of fact [1929, p. 31]."

# CRITICAL THINKING

Critical thinking has seldom been differentiated from problem solving with an identified pedagogical definition or developmental sequence. Usually, it is associated with problem solving and inquiry or it is vaguely assigned to higher levels of the cognitive taxonomy. This has created problems because critical thinking is a learning skill most teachers seek to develop. We posit both that critical thinking is identifiable and sequential, and that in our modern changing world, reasonable and practical planning for critical thinking experiences in the curriculum is vital. In this section, the experiential taxonomy will be used as a theoretical and organizational base for a taxonomic sequence of critical thinking. This base provides a rational and practical way for teachers and curriculum writers to approach critical thinking in planning, implementing, and evaluating the curriculum. Certainly it provides the theoretical framework for critical thinking and a format for the practical application of that framework. We will utilize only the major categories of the experiential taxonomy for a taxonomic sequencing of critical thinking. At each level, we will identify and discuss certain progressive aspects of this process.

Critical thinking is a logical process of interaction and of making choices with given sets of variables and manifests itself taxonomically as this process of interaction and of making choices develops. Included in this taxonomic sequence are the relational teacher role models described in an earlier chapter. In this way, practitioners who want to develop appropriate critical thinking activities within their classroom can use the taxonomic basis and the teacher role definitions more effectively. The taxonomic sequence for critical thinking follows.

## EXPOSURE

At this beginning level, the critical thinking process involves the recognition of variables or of sets of variables within a given context. Learners also begin to recognize that there are differences within the given context and respond to those stimuli presented that point out the existence of such variables. Critical thinking is a process that needs to be carefully planned by the teacher. It is important at this level to prepare an environment where there are recognizable variables. The teacher then motivates the learner to search for those variables. The teacher becomes, in effect, the motivator and manipulator of the contexts and of the setting. If the variables provide sensory stimuli for the learner, this will usually enhance the critical thinking process. The variable, when it can be seen, heard, felt, tasted, or touched, can stimulate the critical thinking and learning process. The teacher should plan for motivating activities that will motivate the learner to respond to the sensory stimuli and to become aware of the variable. The key element at this taxonomic level is the recognition of the variable or variables. The chief learner characteristic is curiosity: Why? is the basic question. This question is similar to that in problem solving, but the purpose is different. Here the object is to investigate, to find out. The reason for continuing the experience is to find out why. A solution is not the purpose of critical thinking but, rather, the satisfaction of curiosity and of the sense of wonder generated by the variables. It is in this context that the difference between critical thinking and problem solving can be seen. The teacher must also motivate the learner to a readiness for further inspection of the variable. When this occurs, the learner moves to another level in the taxonomic sequence of critical thinking, that of participation.

*PARTICIPATION*

This is the data collection and definitional level of critical thinking. When learners attempt to develop definitional constructs about the variable or variables, they may assess the data at hand and determine what data or information are still needed to further define or verify the variable. Here the attempt is being made to discover why there is a variable and what differentiates it from the rest of the environment. There are an accumulation of data and the beginnings of a frame of reference resulting from the verification and definitional process. Thus, the learner begins to note the variable in its broader context and to determine its components. At this level of the critical thinking sequence, the student begins to make inferences and expresses these through tentative manipulations of the variable in relation to other contexts and with other resources. The teacher fosters critical thinking at this taxonomic level by serving as a catalyst in initial student interaction with the variable, by reinforcing positively the need for additional data, and by providing supportive feedback when it is attained. The teacher is a data presenter, a clarifier, an organizer, and an effectuator of further experience. The teacher's role is vital in bringing the learners through this critical thinking process. The data collection and definitional work surrounding the variable help prepare the learners for the next level of the critical thinking process.

## IDENTIFICATION

Following the recognition and definition of the variables, the learner is prepared to test previous perceptions and assumptions about the variables in order to gain a deeper insight into their structure and, thereby, obtain a clearer picture of the organization of the variable and its uses. The mass of acquired data can now be organized more selectively. That which appears valid is clarified and can be identified and expressed in some symbolic manner. During this stage of the critical thinking, the learner applies the data already accumulated to existing situations. The learner makes the application through the associating, the categorizing, and the evaluating processes. Upon completion of these processes, certain identifications have been reinforced and others reduced in strength or eliminated. The learners have, at this stage of critical thinking, a basic structure upon which to build. The teacher's role throughout this stage of critical thinking is that of moderator. The teacher as such must involve the learner not only in discussion, in questioning, in research, in sequencing, and in hypothesizing but also in the recognition that these strategies are vital to the thinking process. Hypothesizing is a key teaching strategy at which learners should become adept as they move through the critical thinking experience. Experimentation becomes a major way for learners to continue through the critical thinking process. The teacher moderates the activities at this time and, thus, fosters a deeper exploration of the variable. The teacher must be adept at discussion, at helping to analyze, at assisting in making choices, and at other similar strategies. These changing teacher activities involves a shift in the role character of the teacher who begins to place the burden for further exploration on the learner. As this occurs and as the learners begin to internalize the critical thinking process, a new taxonomic category in the process becomes apparent.

## INTERNALIZATION

Here the learner becomes involved in generalizing and in trans-fer learning. Based on the activities and results of those activities during the identification stage, the application of the variable to new contexts and situations begins. The learners begin to realize how the variable fits and works and begin to use the generalizations developed at the identification stage in new situations. Actions now evince a behavior modification, a new skill or attitude, and the ability to set that behavior, skill, or attitude to a variety of situations. The variable and its influence are accommodated in the learner's own thinking style as is the internalization process itself. When these evidences of the inter-nalization of behavior of the critical thinking process and of the concomitant behavior occur, the role of the teacher becomes that of a sustainer. The teacher's role is to support the behavioral change, the skill, or the attitude and to help the learner use these in different contexts to further test their utility. The teacher becomes more reac-tive in this role and somewhat less dominant. Strategies of a re-motivation, of praise, and of reinforcement are pertinent at this level of critical thinking. Included are role playing, simulations, skill rein-forcement, comparative–contrastive analysis, and summarization. These strategies help internalize both the critical thinking process and the new behaviors. Throughout this stage no attempt should be made to give the learners a sense of "arrival" or of having completed the continuum of critical thinking. When the learners begin to see critical thinking as a part of the learning continuum, the teacher will know the learners are moving into the final level of the taxonomy of critical thinking, that of dissemination.

## DISSEMINATION

At this point in the critical thinking process, the learners are able and willing to relate to others the impact of the variable on them. This is voluntary and even imperative in terms of the modified behavior and in terms of the process involved. It is the social and sharing context of an internalization of the process. This step should always be voluntary and never should be an obvious imposition of the teacher. Yet the teacher must plan for this and include strategies in any effective curriculum plan. The teacher at this point assumes a more extrinsic role, becoming a constructive critiquor. The role responsibility is to see that the continuum of experience continues and that the learner is challenged to look afresh at the variable in a new set of circumstances, as well as at another variable built from the one just dealt with. The teacher must be sure that new stimuli for critical thinking are available to the learner. In effect, the whole process begins anew for the learner and a new challenge for critical thinking starts again.

Through this taxonomic sequence, one can see how the process of critical thinking develops as the experience moves from exposure to dissemination. The careful and well-prepared teacher will take this sequence into consideration as in making plans to help learners use critical thinking to deal with their environment in new ways. Its implementation will greatly enhance the teaching–learning act in the classroom.

## PROBLEM SOLVING

Problem solving is a pedagogical process basic to the learning process. A classroom or learning setting in which the student has little opportunity to work through the problem solving mode has, in terms of new learning, serious limitations. Problem solving can be, for the student, the impetus for great strides in measured concept growth both in recognized learnings and in personal satisfaction.

Yet problem solving is a process that many teachers neither understand well nor carefully implement. There are many existing definitions of problem solving; some are so esoteric that they have little practicality; others are rooted in practice but have little or no theoretical or philosophical substance. Both approaches merit little in terms of real teacher understanding of the nature of the problem solving.

In this chapter, a taxonomic sequence for problem solving is proposed that combines both theory and practice and that is effective in the classroom. The experiential taxonomy is used as a reference and, again, we will note only the major categories of the process. At each level of the taxonomy, certain progressive aspects of the problem-solving process become apparent. As in the two foregoing areas, differing sets of student and teacher behaviors are involved at each level of the process.

*EXPOSURE*

At this initial taxonomic level, both teacher and student are involved in problem identification. This requires that the teacher provide stimuli and expose the learners to the fact that there is a problem and that as growing individuals they can, and in certain instances must, deal with it. The teacher must help the learner become aware of the problem in order to identify its existence and must then bring about readiness for participation in the problem. Here problem identification is brought about through an arranged environment and through stimuli that require within the learner some reaction of response. The teacher's role at this phase of problem solving is again that of motivator who provides the appropriate setting and stimuli for problem solving. The planned objective for the learner is to identify the problem and to express willingness to pursue it. Teaching strategies appropriate here are demonstration, use of media, directed observation, and interaction–questioning. Once the problem is identified and the learner expresses readiness for continuing participation in the problem-solving process, the next taxonomic dimension to problem solving is available to the learner.

*PARTICIPATION*

At this stage, the learner begins to deal with the problem, to gather data on it, to intellectualize about it, and to define its parameters. At this process level the learner must not only clarify what the real problem is but also develop solution criteria and begin to identify constraints faced. Criteria for solution may be as simple as having to "be in outline form" or as difficult as passing federal audit or receiving Board approval. Constraints most frequently involve time, money, space, personnel, or expertise.

The strength of the teacher's role at this level is apparent. As a catalyst, the teacher must not only bring the learners together with data and resources, thus providing a format for real interaction with the problem, but also ensure that learners perceive the solution requirements and the obstacles to be hurdled. A young learner, for example, may know that some change is to be brought home from a purchase at the store, but unless he is aware that the task calls for subtraction and with the constraint of not knowing how to "borrow," there can be no further progress. The teacher at this process level should include data gathering, imitation, simple questions invoking data recall, estimating and manipulative activities, dramatic play, and discussion. The teacher must continue to stimulate, to react, and to present information in order to clearly define the real parameters of the problem. Solutions are not developed at this point, though tentative assumptions can be a part of the intellectualizing and the data exploration at this level of problem solving. The data exploration and intellectualizing about the problem are ways to define more clearly and specifically the parameters of the problem. When the problem has been clearly outlined and explored in its many facets, the learner can begin to approach resolutions of the problem. When this occurs, another level of problem solving comes into focus at the identification level of the taxonomy.

*IDENTIFICATION*

Until this time, the learners' relations to the problem have been extrinsic in that they have been reacting to the problem and to related resources. At this taxonomic step, the learner's association with the problem-solving process increases in commitment, sometimes unconsciously, and a new dimension of problem solving is begun. The learner identifies and explores optional solutions here. Alternatives are examined and can be tried on a trial-and-error basis with less variable solutions being cast aside. Teaching strategies include research, conferencing, experimenting, hypothesizing, sequencing, and discussing. The teacher's role during this stage of problem solving is that of a moderator between the possible problem solutions and the learner. Although less dominant, the teacher's role does demand student consideration of solution criteria; and possible constraints are noted as they affect the problem solutions. The teacher remains a strong resource leader but encourages learners to pursue for themselves optional problem solutions. Strong questioning strategies are important here as the teacher must be prepared to moderate data and varying optional resolutions to the problem. Whereas problem solving is both an individual and a group skill, in individual situations the learners should be taught the value of interaction with peers as a means of testing their judgments and their hypotheses. The teacher should be proficient at individual and group interaction, at discussion techniques, at conferencing, and at the interactive strategies. During this process of problem solving, optional solutions, as noted above, are discarded and the learners begin to move toward a particular solution to the problem. When selecting particular solutions, another taxonomic level (internalization) of the problem-solving process comes into being.

## INTERNALIZATION

At this level of taxonomic problem solving, the learner pulls together all the pieces in a matrix, actually choosing one of the optional solutions to the problem. In this choice there is both an expansion and a fusion of data, which results in some kind of modified behavior or new action based on the one selected solution. The choice of solution is based on wider and deeper probing into the actual evidence gathered during previous problem-solving activities and on a testing of its viability. The learner is able, upon completion of this level of the problem-solving process, to verbalize or to demonstrate the evidence that led to the choice and resulted in changed behavior. It is the task of the teacher to sustain and to support the choice and the changed behavior—assuming their positive impact on the learners. Teaching strategies appropriate during this process are role playing, simulation, skill reinforcement, comparative–contrastive analysis, higher-level questioning, and summarization. An effective teacher will have planned for some type of behavior modification before the sequence began. A major differentiation from the pre-planning occurs, however, in the experimental problem-solving mode, in which the resolution may not be known to the teacher and depends on thorough data analysis. When this occurs, it is still incumbent on the teacher to plan carefully; the planning centers more on the process of problem solving and less on an anticipated solution. The teacher must be more sure that the process is learned than that a solution is reached. The same taxonomic levels of problem solving remain valid, however. When the learner has accepted a solution, the final taxonomic level of problem solving becomes apparent.

## DISSEMINATION

After internalizing the solution to the problem, the learner implements the solution, initially in the problem situation and subsequently, where feasible, in a variety of contexts. Others are made aware of the solution and of the learner's reaction to it, first as they may be a part of or affected by the problem resolution, and second, as the student attempts to use or to influence others to use the solution in equivalent or even in new situations. Manifested behavioral changes are observable and others may learn from them at this level. This, again, is an extrinsic aspect of problem solving. Problem solving began as an extrinsic stimulus became intrinsic in the identification and internalization stages and now finally reappears at this taxonomic level as manifested operant behavior voluntarily shared with others. It should be remembered that dissemination is voluntary and should be so, both in the process and in the particular problem and solution. The teacher's role again changes and the teacher becomes a constructive critic. Several strategies used at this level are dramatization, laboratory activities, presentation–reporting, group dynamics, and seminar. The teacher's purpose at this time is to stimulate and to expose new contexts in order to keep the cycle going beyond the existing situation. In effect, the teacher is beginning anew the process of problem solving.

These, then, are the taxonomic levels of problem solving. A teacher who can attend to these levels and plan for them in the classroom setting will greatly enhance the learner's opportunity for greater growth in the skills and in the process of problem solving. The teacher who is effective and skilled in these three processes (creativity, critical thinking, and problem solving) and who can promote classroom settings, learning environments, and stimuli to encourage them, will ensure an observable increase in student learning and in interest in the planned classroom activities.

# 8 | Taxonomic Organization of Teaching Strategies

In recent decades there have been numerous studies of variously defined teaching strategies. These studies have been keyed to teaching models and even to extant taxonomies. Sometimes these strategies have been carefully explicated and have had a positive relationship with classroom usage. On other occasions theoretical statements have shown understanding of, or relation to, the real world of teacher–learner interaction in the classroom. In spite of all these studies, there remains a need for consideration of teaching strategies for a successful learning sequence for students. We suggest that there is a need to consider the relationship of specific teaching strategies to the teaching–learning act. We propose also that there is a definite sequence to the teaching–learning act and that a sequence of teaching strategies becomes apparent as one analyzes the process of teaching–learning from the time a learner first becomes exposed to a learning experience through the point at which the learner has internalized the experience and is disseminating it to others. Along this learning experience continuum, 25 basic teaching strategies have been defined and have been sequenced in terms of the experiential taxonomy. These basic teaching strategies are broadly defined and may implement or include several substrategies, techniques, or approaches to the teaching–learning act. There are five major categories to the experiential taxonomy and for each category five teaching strategies are suggested. The experiential taxonomy on which these strategies are based is a gestalt taxonomy providing educators with a framework for total human experience. It is also a functional taxonomy, which teachers and curriculum planners may use as a practical basis for planned learner activities. In utilizing this taxonomy for the sequencing of teaching strategies, we will discuss only the major categories of the taxonomy. This will enable the reader to see the larger flow of experiences and strategies related first at the major categoric levels of that experience and later to a more discriminate study. Educators then should be able to plan taxonomically sequenced teaching strategies in their classrooms as they arrange for their students' learning experiences. We feel that if teachers are aware of the teaching strategy sequence, they can plan more appropriately and teach more effectively. They can further use the teaching strategies and the Experiential Taxonomy Teaching Strategies Coding System as a basis for self-evaluation (see Chapter 9). It should be noted also that a good lesson may not utilize all the strategies at any given taxonomic level and almost probably not the total 25, throughout the experience.

The reader should understand that we have identified only a limited number of strategies. Within each of those, there are specific substrategies and/or related activities. In effect, one should recognize that within each defined teaching strategy there may be a wide variety of activities or of more specific strategies that would serve the same function and that would definitionally refer to the broadly defined strategy named here. For example, demonstration is an identified strategy beginning at the exposure level. A teacher may use demonstration as a means for learner exposure to an experience in the classroom in scores of ways, but all are varieties of demonstration. This is true of many other teaching strategies. The reader should also note carefully the options, the substrategies, the techniques, and the activities implied in the strategy definitions following the chart of teaching strategies descriptors, pp. 112–139. These definitions imply more specific interpretation as teachers apply them in their own classrooms.

A further word on teaching role models should be mentioned. These are described in Chapter 6 and define taxonomically the changing teacher's role in a classroom learning experience as the learning experience occurs through the taxonomy. The role models exemplify the role paradigm a teacher assumes or needs to assume as learning progresses through a planned series of activities in a defined teaching–learning interaction. The reader should refer to Chapter 6 for further explanation of those role models.

In the following, we present the taxonomic sequence of identified teaching strategies. We provide an overview of the teaching strategies keyed to the taxonomic levels. These are defined with brief descriptors that encapsulate the larger definitions and that follow the chart. The chart itself contains the column Teaching Role–Behaviors, which is a brief explanation of the teaching role model and of the behaviors appropriate to each major category of the taxonomy (see Table 8.1).

Before turning to the chart, the relationship between the taxonomic level, teaching role models, and teaching mode needs comment. Readers should keep these in mind as they review the chart.

| Taxonomic level | Teaching role model | Teaching mode |
| --- | --- | --- |
| Exposure | Motivator | Incentive conditioning |
| Participation | Catalyst | Data gathering |
| Identification | Moderator | Research |
| Internalization | Sustainer | Skill reinforcement |
| Dissemination | Critiquor | Presentation–reporting |

The five teaching modes represent a kind of overarching methodology at each taxonomic level. The five strategies at each level of the taxonomy noted on the chart fit into the general rubric of the teaching mode. Obviously, as the learner progresses through the taxonomic learning sequence, the role of the teacher must change. The effective teacher will keep this in mind in working with students.

A study of the chart can be very useful to the teacher for planning and self-evaluation. In coding lessons, we will use this chart as a point of reference. Where one is working with other teachers, this

**Table 8.1**
Taxonomic Organization of Teaching Strategies

| Taxonomic level | Teaching strategy | Descriptor | Teaching role and behavior |
| --- | --- | --- | --- |
| 1.0 Exposure | | | Motivator—presenting and focusing data; demonstration, presenting, and interacting with students; teacher's aid is dominant. Students must be motivated to continue through experience. |
| 1.1 Sensory[a] | 1.1 Incentive conditioning (goal setting) | Introducing an experience, goal setting, establishing stimuli | |
| 1.2 Response | 1.2 Data presentation | Providing information, identifying experience, lecture, use of media | |
| 1.3 Readiness | 1.3 Demonstration | Demonstrate a principle—show how to do, use realia–models | |
| | 1.4 Directed observation | Focusing on particular or selected stimuli, establishing parameters, telling learners what to look for | |
| | 1.5 Data exploration | Interacting with data or with selected stimuli, establishing a readiness for further experience, usually in the interrogative mode, preparation for participation | |

| | | |
|---|---|---|
| 2.0 Participation | | Catalyst—working with the data. Teacher role is vital here to continue student learning. Strategies 2.3 and 2.4 are most helpful in fostering peer interaction and peer teaching. Use groups, large or small, to foster interaction. Peer teaching can be incidental, accidental, and planned. |
| 2.1 Modeling–recall | Mental or physical reproduction (imitation) of a given or known example of concept (model); learner recall based on a known reference | |
| 2.1 Representation | | |
| 2.2 Expanding data bases | Accumulating appropriate resources, generating data, reading, viewing, listening, discussion; "Why" questions or evaluative questions bringing added data | |
| 2.2 Modification | | |
| 2.3 Dramatic play | Unstructured role or situation playing. | |
| 2.4 Manipulative and tactile activities | Use of realia, media, and hands-on activities, how to use materials | |
| 2.5 Ordering | Sequencing of data, arranging data, establishing material hierarchy, defining frame of reference; can be done by teacher to prepare student for next level of taxonomy | |
| 3.0 Identification | | Moderator—Here the teacher's role is to act as resource leader and moderator, helping students become users of data. Field activities are important here as are 3.3 and 3.5 in fostering peer interaction and peer teaching–learning. Once students have a field activity and begin to be users of the data, the teacher's role is to deepen the involvement to the point where oral evaluation can become internalization. |
| 3.1 Field activities | Selection and retrieval of appropriate data, directed field work, reading | |
| 3.1 Reinforcement | | |
| 3.2 Using data | Using, assessing, and interpreting data through observation and experimentation, recording data, explaining | |
| 3.2 Emotional | | |
| 3.3 Discussion–conferencing interaction | Exchanging points of view and interaction relating to field and lab activities—teacher(s)–learner(s), learner(s)–learner(s), questioning, information giving, clarifying information, comments | |

(continued)

**Table 8.1—Continued**

| Taxonomic level | Teaching strategy | Descriptor | Teaching role and behavior |
|---|---|---|---|
| 3.3 Personal ——→ | 3.4 Hypothesizing | Conceiving and using provisional assumptions as the basis for reasoning and action, can introduce as a question, can be stated through a creative expression such as a picture, or a model | Sustainer—As the students begin to express learned behaviors, the teacher sustains and augments the behavior. Strategies 4.2 and 4.3 will help those who may not have learned the material on the first pass to learn from peers. These two strategies are a nonverbal mode of testing and evaluation (3.5). |
| 3.4 Sharing ——→ | 3.5 Testing | Trying conditional hypotheses in multiple situations, asked as a question, confirming hypotheses, applying data | |
| 4.0 Internalization | 4.1 Skill reinforcement | Use of acquired skills in a variety of contexts and ways, questions applied to new context | |
| 4.1 Expansion | 4.2 Re-creation | Re-creation drawn from situations or activities with behavior expectations broadly defined. Sharing, discussing, using internalized behaviors, may even be written | |
| | 4.3 Role play–simulation | Structured role playing, overt demonstration of learned skills and roles—involves social context, affective orientation for role play, simulation cognitively oriented, demonstration | |
| 4.2 Intrinsic | 4.4 Comparative–contrastive analysis | Apply internalized skills to new situations, involves the analysis of similarities and differences in systems or situations, can be in the interrogative mode | |
| | 4.5 Summarization | Review of experience, identification of totality of experienced activities, developing reports, test of skills | |

110

| | | |
|---|---|---|
| 5.0 Dissemination | 5.1 Reporting | A written or graphic report sharing, showing, or explaining an experience |
| 5.1 Informational —can be dissemination level for each strategy | 5.2 Oral presentation | An oral sharing, showing, or teaching the skills, product, or result of an experience, voluntary sharing, showing |
| | 5.3 Dramatization | Formalized presentation of internalized roles |
| 5.2 Homiletic— can be dissemination level for each strategy | 5.4 Group dynamics | The interactive context of dissemination with intent to inform and to influence |
| | 5.5 Seminar | Total participant interaction for sharing ideas, the effect of the experience, and beginning new experiences |
| | | Critiquor—the teacher brings a new context to the students. As critiquor, the teacher's role is to start the cycle of learning anew and to keep the students moving along the learning continuum. |

[a] Please note that those strategies that are linked to subcategories of the taxonomy are shown by arrows. Although the major concern is with a learning progression through the taxonomy, it may be helpful to note how the strategies are related to the subcategories. This also serves to emphasize the hierarchical nature of the strategies. As a lesson progresses and has impact on students, teachers can be assured that they have moved not only sequentially through the taxonomy but also hierarchically through taxonomically keyed teaching strategies. Please note that at their dissemination level any of the strategies can be informational and any can be homiletic. The emphasis is the criteria or the level of dissemination.

chart can, likewise, be invaluable in any discussion of the planning, implementation, and evaluation of learning experiences.

## TEACHING STRATEGY DEFINITIONS

Described later are 25 basic teaching strategies keyed sequentially to the experiential taxonomy. Five strategies are identified for each taxonomic level. Within the definitions of these strategies it should be noted that there may be several substrategies, techniques, or variations of the basic strategy. A teacher should interpret these definitions within the context of individual teaching style and should utilize the defined strategy or any implied substrategies in any way pertinent to the needs of the particular class or individual class members. In the teaching–learning act, flexibility and adaptation are necessary. A teaching strategy is really a combination of preparation and personality; thus the teacher would interpret each learning strategy according to the actual teaching–learning situation. We will define each teaching strategy in its taxonomic sequence. We will name the strategy and follow with the brief descriptor used in the chart and then a more complete definition.

### EXPOSURE

#### INCENTIVE CONDITIONING

This is defined as introducing an experience, goal setting, and establishing stimuli. It is the initial step in any learning sequence. It is the starting point and, when the teacher arranges the environment, provides the stimuli and begins to shape the direction of the learning sequence. This is the classic time for the teacher to be seen as the motivator. The teacher must get things going, manipulate the setting, and provide stimuli for the learner. The effective teacher utilizes a number of incentive conditioning stimuli, using media, posing a problem, asking a question, presenting a puzzle, or making a statement. The most important task is to stimulate in the learner a desire, indeed a felt need, to pursue the experience through the achievement of the goals and the objectives. Another task is to set goals or to give the learner some idea of direction. This can be done in a variety of ways ranging from teacher presentation to learner–teacher interaction. It should be kept in mind that an introduction implies the content, the sequence, and the parameters of the whole experience. Consideration of this point is paramount as student learning is planned. Incentive conditioning ranges from what Gagné (1969) calls signal learning and Skinner (1968) defines as operant learning to a chaining of the responses to further activities along the continuum of the experience. In the experiential continuum, the response at the exposure category is not the end but rather the beginning of the experience. Incentive conditioning needs a motivational response from the learner that, when linked to other aspects of the environment and to the learner's store of remembered experiences, produces an internal need to respond further to the experience. The burden of incentive conditioning and the selection of stimuli is upon the teacher, who must select those stimuli that will provide the links to the elements of the experience.

There are two basic forms of stimuli that the teacher must consider when selecting the stimulus for incentive conditioning. First, there are the fixed stimuli that are basically components of the arranged environment. They tend to remain somewhat static in the initial setting, though as the experience develops they can change. They are used to point toward and to augment the second kind, which are the operational stimuli. Operational stimuli are those specifically used by the teacher to provide impetus for the potential within the ongoing experience. Choice of these operational stimuli is, of course, the instructional prerogative of the teacher and can include a whole spectrum of stimuli from concrete to abstract. They should be chosen for the purpose of eliciting a sensory response from the learner. Imperative in the selection of stimuli is that the teacher know the kinds of responses anticipated or permitted in the ongoing experience. An awareness of the individual differences within the learner group coupled with a consideration of the possible responses by individual learners according to their learning styles, modes, and preferences are necessary to enable any teacher to use those responses that will move the students along the experience continuum.

DATA PRESENTATION

This involves providing information, identifying an experience, lecturing, the use of media, or any method of presenting data. In any learning situation, the student needs a data base within which to operate. This teaching strategy provides that base. The student has been introduced at this point to the experience and has some construct of the goal or direction of the experience. What is needed now are some data or information in order to continue. The role of the teacher is still that of motivator, so the data presentation needs to be accomplished in as motivating a manner as possible. The teacher must bring data to the learner so that enough information can be presented about the experience to motivate the learner to discover its potential and parameters. With data presentation, at this point, the learner receives information rather than exploring it. It is a response to an exhibited learner need to gain more information; the learner uses the data about an experience to respond to it internally and as inspiration to continue with the experience. The teacher acts as a continuing motivator whose task it is to help the learner make the decision to continue with the experience. The teacher brings about further motivation by providing the student with resources to review and with data for learning. It is essential that the teacher be thoroughly acquainted with the content discipline in order to bring the appropriate resource to the learner. The teacher must also be cognizant of ability levels, so that the materials meet the need of the learners at their achievement levels. The age, the experience, and the maturity level of the learner will be factors in selecting which data are appropriate and what the most effective mode of presentation might be. In every case, however, there needs to be as wide a range of presented data as possible.

When media are used for data presentation, the wide spectrum of materials ranging from print media to audiovisual media should be considered. Although media are used at every taxonomic level, their purposes vary as experiences develop. At the exposure level, the use of media must be judged from its motivational value and, therefore, greater attention given to its uniqueness. At later experiential levels, use of media can become a teaching substrategy that relates to other basic strategies such as research, skill reinforcement, and comparative–contrastive analysis. At the exposure level, it is a definite strategy involving the teacher in a motivating teaching style. The environment is manipulated and stimuli including media introduced. Media themselves can be very provocative and evocative stimuli if used effectively. Several elements enter into their use as a strategy. These include (*a*) definition of expected outcomes; (*b*) choice of materials; (*c*) preparation of learners; and (*d*) evaluation and postusage planning. Defining expected outcomes is a first step without which there is little or no point in continuing a particular experience. Once they are defined then choice of materials becomes important. Those that are pertinent are selected. Too often extraneous media are used simply because they are available and not because they are the best available or are appropriate to the learning experience. When this happens, learning is compromised. Media, if they are to be used, must both interest learners and move them toward a developing experience. Third, learners must be prepared by using media or they are no more effective than the nightly telecast. They need to have a purpose for using the materials and should know what they are to look for or listen to. This, then, provides a purposeful use of media and enables the learner to move on to other categories of the learning experience. Implicit in the use of media is the need for review or preview of the material by the teacher. The teacher should note the points to be emphasized and how best to do this in the class setting. Finally, the teacher should review with the learners what they saw and heard and should know what their perceptions of the media were in terms of the planned experience. In this review, the teacher can determine the relationship of this use of media to other motivating teaching strategies such as directed observation, stimulus–response, and the technique of interaction–questioning. Once again, it is emphasized that the teacher's role in the use of media at this taxonomic level is to move learners to a new level of experience, more specifically, to the level of participation.

DEMONSTRATION

This involves showing how to do, demonstrating a principle, and the use of realia–models. This strategy does what the name implies. Here the instructor, whether teacher, resource leader, or learner, is showing what something is and how it functions or how to perform a skill, an art, or a process. It is yet another teaching strategy designed to motivate student interest and attention. It can be carried out in either of two methods: the expository (explaining) method or the inquiry (problem solving) method. In the expository method, the teacher or resource leader is attempting to show the learners something they will need for future development of the experience. With this method, the learner response is more intrinsic and individual than in a class or group context. Though the purpose is to motivate, the instructional style is often didactic. The purpose is to show, to explain, and to inform. With the inquiry method, the teaching style is interrogative and is designed to pique interest and to develop both extrinsic or group interaction as well as individual learner response. The inquiry style or method, like the expository style, anticipates an emerging need within the learner to move to a further and deepening experience. Both teaching approaches in this demonstration strategy depend on the learners' sensory reaction. The leader can plan demonstrations for learners with visual, auditory, kinesthetic, or other learning styles or preferences. Demonstration is enhanced if learners themselves are able to manipulate, explore, or otherwise deal with the object, the solution, or the idea being demonstrated. In using demonstration, the teacher needs to provide a framework of response within the learner by allowing for time for demonstration and reaction. From learner reactions and responses, the teacher can develop questioning strategies or interaction strategies to advance the learner to further levels of experience and, more particularly, to participation. Demonstration can be an important strategy in advancing a learning experience through the taxonomy. By the very nature of demonstration, it is the teacher or demonstrator who manipulates the environment for the learner and who provides the stimulus for learner response.

DIRECTED OBSERVATION

Here the teacher is focusing on particular stimuli, establishing parameters, and telling learners what to look for. This teaching strategy again places the teacher in the role of motivator and manipulator. The teacher now is directing learner attention to stimuli and to the particularized aspects to which the student is expected to react. In effect, the teacher is narrowing the kind of response to the stimulus or, to put it in another context, is organizing the parameters of the response. The learner knows not only what to look or to listen for when the identified stimulus is shown but also to which particular facets of the stimulus to react. Two elements are involved. First, the teacher can identify the expected learner response and at evaluation find out if the students responded in the same way. Second, the teacher can particularize the stimulus without giving an anticipated response. At evaluation, the learners can then discuss similar and different responses. Both of these approaches to directed observation are valid for classroom use and have a place in the teacher's repertoire. Directed observation is perhaps a limiting term because it implies the use of only one of the human senses. Perhaps directed sensory response would be more appropriate since it suggests the use of all five human senses. Directed observation as used in this section means directed visual observation, directed auditory response, directed olfactory response, and so on through the senses. Directed reading can also be a part of this learning strategy when the reading is to motivate student exposure to an experience. The point of the strategy is that the learner is being directed toward responses that will lead to further levels of experience and will to a range of expected internalized behaviors. The teacher focuses and directs the learner toward an avenue of experience that leads to objectives of internalized and disseminated behaviors.

DATA EXPLORATION

This involves interacting with data or with selected stimuli, establishing a readiness for further experience, usually in the interrogative mode, and preparation for participation. This teaching strategy is one that depends on teacher–learner and learner–learner interaction. It is primarily interrogative, with a twofold purpose. First, it gives the teacher and the learner the opportunity to get into the data presented and to begin to visualize their parameters and their possibilities. Second, it is at this point that learners and teachers make mutual but individual decisions as to whether the experience is important enough to continue through the taxonomy. Thus, experiences are abandoned or entered into with new psychological and intellectual thrusts. The interrogative teaching mode is most pertinent at this step because it is in this manner that the teacher and the learner can effectively ruminate about their exploration of data. The teacher's role is still strongly that of motivator; but at the same time the teacher must be sensitive to the learners readiness to move on to the next levels of the taxonomy. In essence, it is through this interrogative exploration that the learner moves to the participation level. Several studies of classroom questioning have been most helpful in analyzing the interrogative teaching mode, particularly the work of Sanders (1966), who based his work on Bloom's (1964) cognitive taxonomy. His incisive interpretation of Bloom permitted the development of a hierarchical sequence of questioning, beginning with memory and ending with evaluation. This work contributed greatly to the quality of classroom questions. Even so, there is another dimension to this strategy, namely, that of interaction between teachers and learners and among learners. Interaction is really communication and in common language is today a somewhat pandered word. In this context, however, it means purposed dialogue, including questions and responses, that prepare a learner to participate in further aspects of the experience. Data exploration, as noted earlier, is a twofold strategy. The first part of the strategy is a carefully planned sequence of questioning that is designed to lead toward preplanned responses and that readies the learners for participation. Data exploration in this context becomes more effective as the carefully structured question sequence, leads inexorably to the planned readiness and participation. Questions should be logical, linked, and sequential; the learner will then respond with a readiness to participate. If this were all there was to data exploration, there would be little problem. Human beings, though, do not always give the "right answers" or respond logically. It is here that interaction comes in. Learners may react to a question with personal responses or with existing behaviors. The teacher must then interact deliberately either to bring the learner back on the track toward planned readiness or to adapt the learner response to the planned objective. This strategy requires an alert teacher who has planned the learning experience carefully and who possesses the interaction skills to bring the learner back toward the plan for further learning. This must be done carefully so that the learner will continue to be motivated to participate in the experience. When the planned sequential questioning and the focusing teacher–learner interaction here worked together, the learner is ready for participation in the experience. The teacher's role as motivator is urgent at this step because decisions need to be made about the direction of the experience. The teacher must prepare for this strategy with great care to ensure that the learner makes the transition to participation effectively.

*PARTICIPATION*

MODELING–RECALL

The learner practices mental or physical reproduction (imitation) of a given or known example or concept (model). Learner *recall* is based on a known reference or on past experience. The transition to the participation level of an experience is not always an obvious or an easily identifiable change. It is, rather, a deepening of personal participation. In this strategy the interrogative mode mentioned previously may well continue and the stage may, outwardly, be hardly distinguishable from data exploration. Yet the quality changes in that the learner begins to participate with the data by recalling past references to the data, by dealing with the data in a definite context, and by being able to relate that context to a perceived model of the experience. At this point in the experiential taxonomy, learners become active participants in the experience. They work with the data they have gathered and deal with it both mentally and physically. They begin to relate it to their past experience and to make modifications appropriate to their own past experience. One way in which this can be done is through modeling and recalling. Modeling here has two connotations. First, the media or material can serve as the model with the modeling being somewhat more abstract than when a teacher, resource leader, or peer does it. This can be a kind of private or covert "walk-through" of the model. Second, the teacher, the resource leader, or a peer can model certain roles or behaviors that can then be imitated by the learner. Recall, likewise, has two connotations. It can be a covert or mental imitation that may not be physically observable or it can be an overt recall in terms of small or large group interaction in the classroom or on the playground. It is the teacher's role to prepare conditions so that learners can utilize both types of modeling and can experience both kinds of imitation. Covert recall almost always precedes overt recall. There must be some inner coming together of past experience with the observed model before there is an overt recall and sharing of modeled behavior. Sometimes this is done almost instantaneously but usually the learner needs time to consider. It is a time for private and personal thinking. The teacher needs to be alert to this need for quiet consideration. Sometimes, feeling the need for measurable student achievement, a teacher may not provide for this necessary kind of thoughtful consideration. Yet it is an essential and integral part of human experience. In this strategy, the teacher remains the controller of the learning environment but acts as the catalyst who brings these experience opportunities to the learners. This strategy, like the others, must be planned very carefully. The teacher needs to know what should be modeled and to select these with care. Data are necessary to this strategy and the teacher needs to be certain learners have had both the data presented and an opportunity to explore the data. Time must be provided for learner rumination and consideration of the data and model. Overt modeling–recall can well be provided for by the teacher. This strategy can lead the learner toward further levels of taxonomic experience, particularly toward the need to expand data bases.

EXPANDING DATA BASES

Different methods include accumulating appropriate resources, generating data, reading, viewing, listening, discussion. "Why" questions are asked as are evaluative questions in order to add data to the experience. At this level of experience, a learner is beginning to broaden the construct of the experience. The exploration of data, the relating of those data to prior experience models, and the recall of those models generate learner need for additional data. It is at this point that the learner feels most keenly the need for more information and for more usable data. The catalytic role of the teacher is most important here. The task is to focus the learner's need for more data and to provide material and media to satisfy that need. The teacher knows data sources, appropriate materials, and constructive medius. As catalyst, the teacher encourages and directs learner participation in the generation of data. Through the use of modeling–recall (2.1), the learners can be a part of the data gathering process. The teacher may also need to reach back to data presentation (1.2) and directed observation (1.3) or to some other strategy to augment the data gathering. The experience of expanding data bases can be very stimulating for teacher and learner and, if appropriately done, should move students to further levels of the experience. At this point in the taxonomy, the learners begin to make modifications in the experience as they layer new data on their initial representation of the experience. It should also be remembered that the learners are not yet in a research mode, which comes at the identification (3.0) level of the taxonomy, but rather are gathering new data and are exploring it. They know enough to make decisions as to what kinds of data they will need, specifically, but do not have enough yet to begin to utilize it as a response to identified needs. They still need to participate with the data in related strategies before moving on to the identification level. This is done in the next three strategies.

DRAMATIC PLAY

This is unstructured role or situation playing. This strategy, although not used as often as it should be in classrooms, can be one of the most dynamic strategies in moving students through a complete experience. Dramatic play can be defined succinctly as unstructured role or situation playing. There are three vital steps in dramatic play. The first step is planning, the second is the activity, and the third is evaluation. Two levels of planning are important, teacher planning and student planning. Basic to teacher planning is the establishment of the need for dramatic play. This need is contingent on the strategies used at the exposure level of the experience and on participation with the data and with other participatory strategies the teacher may have used. The teacher needs to review these first and then to plan specifics such as expected outcomes, realia needed for the activity, what planning needs to be done with students, and how dramatic play will move the learners to identification. Planning with the learners begins with "Let's be the people" and then moves to "What activities will they be doing?" The activities are listed on the chalkboard so that they can be seen. Next, the teacher lets the learners choose what or who they want to do or be. After these decisions, places are assigned and learners play their chosen role. The activity itself is probably the least important of the three phases of dramatic play. It can also be the briefest. Learners need not even know exactly what to do; they need only play the role the way they interpret it at the time. The teacher must, however, be very observant during the activity—and of what the learners are doing or not doing. The teacher may want to take notes for purposes of later group discussion of the activity. Although it may be important to get as many students as possible to participate, total learner involvement is not absolutely necessary. The teacher, as catalyst, provides the physical and/or visual impetus for student involvement.

Evaluation is the crucial element of dramatic play. Questions like "Who did you notice was doing a very good job?" can start the process. Those students who really put themselves into their role are commended and the discussion moves to three questions: "What do we need to learn?" "What do we need to make?" and "What do we need to do?" Each of these should be keyed to the supposition that the activity will be repeated and that the answers should help make the next performance more realistic. Through interaction–questioning, three lists are made. The first deals with what the students need to know, the next is a list of things to make, and the third is a list of more realistic ways to act or to behave. All lists suggest improvements in the dramatic play the next time it is done. The positive quality of the lists allow the teacher and the learners to establish identity with the experience and to move to research and to the improvement of the activities. As the learners reach higher levels of the taxonomy, dramatic play may become more structured role playing, simulation, or dramatization as they internalize roles and objectives. With respect to this strategy a final point needs to be made. The older and more sophisticated learners become, the more sophisticated the strategy itself needs to be. Simple adaptation of the basic strategy can accomplish this. One can, for example, cite a problem and then have students play roles. Shaftel and Shaftel (1967) give a sound explanation of how this can be done.

It should also be noted that other role situations arising from student interaction are pertinent. Learners should be given the opportunity to build for themselves the imaginative roles related to the learning experience. Teachers need both to plan for dramatic play and to provide opportunities for it on an individual and/or small group informal basis. In this context, the same steps of planning, activity, and evaluation are as important as in the large group session previously discussed.

## MANIPULATIVE AND TACTILE ACTIVITIES

Learners use material, realia, and media. Hands-on activities and material use are discussed. This strategy involves that wide range of activities that can perhaps be best defined as "hands-on" activities. In this strategy the learner works with media and material, and activities include everything from paper and pencil tasks to working with clay, with art media or music media, or with construction materials. Manipulative and tactile activities are strategies that persist through internalization. At this participation level, they involve discovering how to use materials, getting used to the feel of tools, experimenting with clay or chalk, and trying something to see what happens. These activities are necessary to enable the learner to discover the parameters of and to foresee the potential within the experience. The teacher can accomplish the objectives in this strategy in a formal or informal mode of presentation. In the formal mode, the learner follows a step-by-step sequence involving manipulative and tactile activities organized by the teacher. During these activities, there is, of course, an interaction–questioning component. This carries over to evaluation, where the teacher needs to find out the results of the activity. In the informal mode the teacher has the material and/or media available, perhaps as part of an instructional packet, and the learner at appropriate times will experiment with, use, and find out about them. These are available in as interesting a way as possible to allow for learner exploration. Evaluation is just as important in the informal mode but is more apt to be done individually or in small group situations. As higher levels of the taxonomy are reached, manipulative and tactile activities will result in creative written works, in art expression, in construction, and in many other ways. This strategy is most appropriate at this level and continues in effectiveness through the identification and internalization categories of the taxonomy.

ORDERING

This involves sequencing of data, arranging data, establishing a material hierarchy, and defining frames of reference. The teacher can do this to prepare the student for the next level of the taxonomy. Ordering is a teaching strategy that greatly assists the learner advancing to identification of an experience. It should come at a time when the learner has been exposed to an experience and has explored its parameters through participation. This is the time when the learner reflects back on the experience thus far, reviews the data, and begins ordering activities to continue the experience. Three elements need to be considered before the learner can successfully accomplish this strategy and move to identification. First, as previously intimated, the learner must have explored the parameters of the experience through data presentation, through data exploration and gathering, through dramatic play, and through manipulative and tactile activities. Second, the learner must have dealt with the data in terms of past experience. Third, the learner must be willing and able to project a sequence of activities into the future. Ordering is, therefore, a futuristic teaching strategy. It involves bringing past experience together with present data for the purpose of projecting possible future action. During this process, an emotional and intellectual identification with the experience begins to take place and there is a transition from participation to identification. The teacher, in this strategy, plays the important role of initiating the experience and helping to organize the reflection, the review, and the ordering. As the strategy advances and continues, by the very nature of the experiential continuum, the student begins to take more initiative. In the final analysis, it is the learner who makes the decision to identify with the experience and to move on through the continuum. The teacher's role is to help the learner logically project the experience continuum to its fruition as identified experience and ultimately to internalized behavior and to dissemination of that behavior. The teacher must be sensitive to the learner's emotional and intellectual set because the teaching role at this level of the taxonomy undergoes a major shift and requires the teacher to move from the role of active catalyst to a more passive yet equally vital role as a moderator. Sensitivity to the learner's place in the experience is particularly necessary here. Frequently, teachers are not able to make this transition and, therefore, it is at this level of the taxonomy that the experience fails to progress to its planned conclusion. The student may later internalize the data, but the process is slowed and the outcome may vary should the teacher fail to make the appropriate role change. The teacher's tasks in this strategy are to prepare data, to visualize what the outcomes of the sequence can or will be, and to bring the learner to a place where the continuum of experience can be foreseen. The learner, then, can project a possible sequence of activities. The teacher, through interaction and questioning techniques, plays a vital role in this projection and at the same time plays an exceptionally important role in moving the learner along the way to identification. Any lesson or learning sequence must reach this level of taxonomic experience to have a vital effect on the learner and to achieve sound pedagogical success. It should also be pointed out that a lesson can terminate at this point and that ordering can be a summary strategy reflecting back on what has gone before and projecting further activities.

*IDENTIFICATION*

FIELD ACTIVITIES

These include the selection and retrieval of appropriate data, directed field work, and reading. At this level, the learner has begun an emotional and intellectual identification with the experience and has *chosen* to progress to additional levels of experience. In the process of ordering, the student and teacher have identified directions and focus for further progress through the experience. Now the student is ready to *do* something. That need to *do* is met through what are termed field activities. A field activity as the term is used here is more than a field trip. It is, rather, any activity within or outside the classroom in which the learner's purpose is to select and retrieve data appropriate to needs established at the ordering stage. Thus, reading, construction, art, working with models, and of course field trips fall into this category. The three basic elements to any field activity are preparation and planning, the field activity itself, and learner response to it. Preparation and planning involve teacher, learner, and the field resource (books, other printed material, realia, media, and known resources). The teacher must have a rationale for the field work, must know how it relates to the total experience, must make provisions for supplying needed data to learners, and must prepare learners to do, to perform, or to respond to various aspects or stimuli in the field activity. Learners, too, should be aware of the objectives of the field activity and, wherever possible, should help plan them. Another aspect of the planning–preparation element is logistical. Availability of media and materials is as much a part of the logistical planning as are transportation, supervision, and permission problems if one plans to leave the classroom for the field activity. Furthermore, the necessary information about the purposes and the expected outcomes should be given to resource people in the field activities so they may participate as fully as possible in helping learners achieve the objectives.

While engaged in field tasks, learners should be able to identify their activity and how it leads them toward the purposes and objectives of the activity. Before they begin, the students should know these required activities and ways of achieving them. The teacher should note any variation from the activities so that the differences can be dealt with on the spot or later incorporated within the context of the experience continuum. Field activities should be active, data selecting and involvement kinds of activities whenever possible.

The third element of field activities, the response–review–evaluation, begins with learner knowledge of the objectives, and continues through learner execution of the activities and assessment of performance as well as the teacher's evaluation of achievement. Learners should be made to realize that for any field activity there is a need to act and to evaluate. The purpose of this stage is to solidify the learner identity with the experience so that the next levels of the experience can be developed and accomplished. Field activities have little efficacy unless they lead to deeper identification and further internalization. The response to the field experience can be manifested in many ways from written response to discussion and creative expression. One can respond to such an experience through role playing, simulation, processing data, and evaluative discussion of the activities. These responses can prepare the way for hypothesizing, for the testing of hypotheses, and for the internalization of objectives.

## USING DATA

This covers using, assessing, and interpreting data through observation and experimentation, recording data, and explaining. The field activities focused on the selection and the retrieval of data appropriate to established experience needs. As the data are selected and/or retrieved they then become the basis for another strategy of identification, that of using data. Learners begin by recording the selected data in clusters, in categories, and/or in usable sets. The next step is to be able to explain the selection of and the categorization of the data. Here the learners should know why they are dealing with the data in certain ways and what some of the possible outcomes may be. Learners also may be directed to observe or even to experiment with selected data. This strategy also involves a variety of laboratory activities. Furthermore, learners at this level of experience will begin to interpret the data into new sets, clusters, and categories. The teacher's role is to foster the interaction of the learner with the selected data and to moderate between the learner and the data. This role has a twofold responsibility. First, teachers must provide the learners with the content and achievement skills necessary to objectives in data use. Second, they must allow the learners to assume responsible direction and freedom of decision. The teacher's role is that of a moderator who helps the learner negotiate the new situations and the new problems arising from the selection and the use of additional data. We have noted that the learner must be able to give the rationale for the use of data. It should therefore be assumed that this process of using data is not done only as an isolated and individual effort but that, rather, it can and often should become a cooperative, interactive, and investigative process in which there is peer interaction about the data in small or large groups. As the interaction continues, it shades off into another strategy, that of discussion–conferencing interaction.

DISCUSSION–CONFERENCING INTERACTION

Here the participants exchange points of view and interact around field and lab activities. The interaction is between the teacher and the learners and among learners, with questioning, information giving, information clarifying, and comments. When one identifies with an experience, there is a felt need to share this identification with peers and with others and to see it in an understandable context. An important teaching strategy to provide an opportunity for this sharing is discussion–conferencing interaction. This strategy also illuminates sequencing and other strategies at the identification level of the taxonomy. Discussion–conferencing interaction means a sharing of ideas, of data, and of findings. It serves two purposes. First, it allows the learners to report what they know and to indicate where they are on the experiential level. Second, it can cause the learners to progress experientially in that they can get new ideas and additional data from others and can share their own ideas. Discussion–conferencing interaction can take place in pairs, in committees, in small groups, or in classroom-size settings. The teacher moderates discussions by asking the appropriate questions, by keeping outcomes in mind, by helping the learners interact positively, and by assisting the learners to establish a means for sequencing and sharing.

Discussion–conferencing interaction needs careful teacher preparation. At this stage it has to be more a science than an art. Conditions need to be arranged for it. Timing is important and the teacher must be aware of the data and of the taxonomic level of the student. As moderator, the teacher focuses on maintaining the direction already established. Therefore, teacher observation of students and knowledge of student progress in the experience continuum is vital. The purposes of discussion at this level are to share, to continue to gather information, and to provide impetus for continuing to the next experiential level. This requires an alert and capable teacher who knows where the learners are in terms of the experience and who can sense when learner identification with the activities of the experience occurs. Conferencing is a logical step involving small and even large groups. It fills the student need for sharing and for group work when groups or individual learners may have worked on different aspects of the same data. This total strategy, which includes committee work, teacher–learner conference, group research, panel reporting, and informal group or individual discussion as well as one-to-one discussion, begins with and fulfills the need for interaction, as the learner identifies with the experience and demonstrates an overt need for interaction with others. It continues through conferencing to meet needs felt in data gathering and research. Finally, the learner expresses a need to tell others of results; of findings; and of the personal, emotional, and intellectual commitment to the experience. The teacher, as an arranger, a stage setter, and a moderator, senses needs, arranges opportunities for conferences, provides the format and context for them, and keeps the interaction moving. The learner bears the responsibility for success; the teacher sets the conditions for success. The teacher should identify and the learner accomplish the goals and expected outcomes for this strategy. The teacher is a kind of ex officio member of each group or conference and acts in that role. The teacher must also be a good listener who notes the interaction and helps moderate it toward planned goals and objectives. Discussion–conferencing interaction continues as a strategy through internalization and manifests itself in the group dynamics at the dissemination stage of the taxonomy.

HYPOTHESIZING

This involves conceiving and using provisional assumptions as the basis for reasoning and action. It can be introduced through a question and can be stated through creative expression such as a picture or a model. A learner who has explored the parameters of an experience, has identified with it emotionally and intellectually, has dealt with it in a research context, and has interacted with others about it can begin to hypothesize about it. Hypothesizing is usually the result of an instructional setting where it is anticipated and planned for. By definition, hypothesizing is the conceiving and the use of provisionally accepted assumptions as the basis for reasoning and action. It is a logical extension of research. Hypothesizing is a necessary preliminary step to internalization in that one posits possible actions or possible solutions based on preliminary data input, on study, and on research. It can, for example, develop from conferencing and from data-gathering strategies. Hypothesizing can be as simple as a speculative "what if . . ." or as complex as the identification, the analysis and the determination of the covariance of specific variables. It can, furthermore, be done as a creative expression. A picture, a diagram, a model, or a composition can be a hypothesis as viable for testing as an oral or written hypothesis. We stated earlier that teachers need to expect hypothesizing and to set the conditions for it. They need to have followed very closely the experiences at the exposure and participation levels and need to be aware that the learner has identified with the experience. For effectiveness, teachers in this strategy need to have looked at many possible variables within the experience and need to be able to bring them to the students' awareness. They must, through learner–teacher and learner–learner interaction, help learners follow the implications of the hypothesis. In effect, the steps are to present possible hypotheses, to investigate them, to incorporate suggestions of others, to try the resulting hypotheses, and then to recognize possible implications and to follow them as far as possible. When these steps are accomplished, the learner has taken a long step toward internalization.

TESTING

Here the learner tries out conditional hypotheses in multiple situations. The teacher can ask a question that will make the student confirm the hypothesis and apply data. It is at this point in the experience continuum that one tries the hypotheses developed and accepts or rejects them on the basis of a review of data and/or of prior and present experience. The learner has come up with some ideas, constructs, or products that are personal hypotheses. At this point, through interaction, review, sharing, and discussion, the teacher and the learners test the hypotheses. Interaction is imperative here as both the teacher and the learners can do the testing. If a creative expression is to be tested as the statement of an hypotheses, it should be discussed in terms of whether it is an accurate representation of data, of skills, or of attitudes related to the individual use of data. Testing can also involve a teacher probe through quizzes or examinations of where a learner is in terms of the posited hypotheses. If the experience continuum is carefully planned, the teacher may already have anticipated possible hypotheses and may have prepared the testing process in advance. On other occasions, learners will develop viable hypotheses other than those that the teacher may have considered.

The process of testing then would need to be developed concurrently with the development of the hypotheses. Testing can also demonstrate the accomplishment of certain skills. At this level, hypotheses are developed and tested but not yet internalized by reinforcement replication and other internalization strategies. It is through this testing of hypotheses that the student moves into the internalization level of the taxonomy. Regardless of how it is developed, this becomes a pertinent and necessary step each learner must accomplish while moving through an experience. It is here that the learner realizes the personal accomplishment and the acquired skills. Now, having attained the projected goals, the learner is ready to internalize this body of knowledge.

## INTERNALIZATION

### SKILL REINFORCEMENT

Here, the learner uses acquired skills and learnings in a variety of contexts and ways. The questions and the hypotheses are applied to new contexts. This is a major teaching mode at the internalization stage. At this taxonomic level, the learner demonstrates internalized behavior through the reinforcement of learned skills, of tested hypotheses, and of creative expressions. Skill reinforcement involves the use of acquired skills in a variety of contexts and in a variety of ways. Learners are able to apply their newly learned behaviors to different sets of activities, increasing their skills in many directions. Sometimes they can express their skills creatively, sometimes through problem solving, and sometimes in critical thinking. The teacher will find it essential that learners have the opportunity to reinforce and to expand learned skills at this point in the experiential continuum. It is not enough to do something once; rather, the learner must repeat it and use it in as many ways and in as many contexts as possible. Skill reinforcement is more than applying the same process to a series of problems, as in a math worksheet; it is applying the process in a variety of ways. It is the teacher's role to find creative ways for the learner to use newly developed skills. The teacher must plan for and implement a whole spectrum of opportunities for skill reinforcement, ranging from drill to creative expression, in order to sustain the learning. Students often need to be made aware that they have acquired new skills, in order to be motivated to use them. Learners become proud of their accomplishments; so praise and positive feedback are important to support and to sustain the new behavior. The role of the teacher is to provide approval and security, and to sustain the learners in their new skills. In effect, the teacher does two things. First, the teacher provides the learner with the opportunity to demonstrate skill in further contexts; and second, the teacher approves of the accomplishment of that achievement. The teacher, therefore, is not just the approver but is also a sustainer of the behavior. Skill reinforcement can be done in any setting. The effective teacher will recognize its important place in the experiential continuum and will use it in its most appropriate manner.

RE-CREATION

Re-creating takes place in situations or activities in which behavior expectations are broadly defined, such as sharing, discussing, or using internalized behaviors. The activity may even be written. Here the learner is asked to projectively re-create or reproduce in thought, in verbal expression, or in written communication, a learning experience. This is the point at which the learner reinforces the learned skills without specific direction from the teacher. The teacher's role is to create a climate for and an opportunity to accomplish this re-creation. Art, music, and creative writing are means by which learners can reproduce a learned experience. This is a time for the teacher to encourage creative expression. A mural or collage, a creative story, or an imaginative report are examples of the way students give new life to an experience. It involves an expansion of the internalization of behavior. The learner becomes more familiar with and more able to use the acquired learnings. Sharing and discussion continue in importance, and opportunity for this verbal aspect of re-creation should be included. Through this strategy the teacher can bring learners who may not yet have internalized the skills to the level of the other students. Re-creation, even though it implies sharing and involves a discussion, is primarily an individual expression of the internalization of skills. It can grow directly out of the testing of hypotheses (3.5) or it can be built around the skill reinforcement strategy (4.1). Its major advantage lies in that it provides for both the expansion and fusion of experience.

ROLE PLAYING–SIMULATION

This involves structured role playing, the overt demonstration of learned skills, and the interpretation of roles. It takes place in a social context. There is an affective orientation for role playing, whereas simulation is cognitively oriented. In all these, the teacher may demonstrate. Role playing becomes an important and even a necessary teaching strategy when learners have internalized the behaviors planned for the experience. Role playing is an affective expression of internalized interpretations. It differs from the dramatic play that is used at the participation level in that insight into the roles played are much more advanced and sophisticated. The learner now has the benefit of more data augmented by research, by sequencing, by discussion, and by testing. Furthermore, the learner has made an emotional and an intellectual commitment to the experience and has learned new skills, attitudes, and behaviors. Finally, role playing is a public expression of internalized interpretations and behaviors, demonstrating that the learner has adopted the new learning, has internalized certain new functions, and can express these through role playing. Role playing done correctly is an overt and outward manifestation of newly internalized behaviors. Learners usually enjoy these opportunities to display new behaviors since they allow a wide variety of expressions. These signalings of internalized interpretations can be revealed formally in reporting form, or informally, in the context of the regular activities. Role playing, as noted previously, is more advanced than the earlier expression at participation. The learner understands the role and, since he has internalized behaviors, the role playing becomes a personal expression of learning. Always, however, a teacher will plan for learning opportunities in this activity. The role playing experience should be carefully planned to allow for the expression of learner knowledge and of an understanding of the role. The learners should be able to demonstrate awareness of what they have done, of how they have done it, and of what its implications may be. The teacher also has a major role in evaluating the role playing. These evaluations often involve oral interaction and discussion between teachers and learners, but these are subject to control of the participant and may not reveal any more than what that individual wishes to impart. The teacher should also consider anonymous, open-ended, projective devices.

Simulations are cognitive expressions of presumed real-life situations. They are best used to move learners beyond the initial learning experience at the identification stage. Learners at this level of experience can simulate particular roles, sequences of activities, and problem situations. They can demonstrate a concept or certain learned skills. Through their simulation, they manifest those behaviors contingent upon a particular role, a set of activities, or a problem. Simulation differs from role playing in that the learner is replicating identified situations or activities with the behavior expectancies and/or activities that are structured in nature and are carefully defined. Many simulations are games or involve gaming. Others are simulations of particularized roles such as those of the people within a city government or in business. In either instance, the student understands the problem or the situation and the expected behaviors involved, can replicate those behaviors, and can evaluate or discuss the behaviors. The teacher here has responsibility for reviewing the simulation, for checking on how it fits into the total scope of the experience, for relating the learner simulation to the total experience, and for helping the learner to evaluate the simulation so that it can be completed through the dissemination level of the taxonomy. The teacher must observe the simulation carefully to be able to verify internalization of planned behavioral objectives. Simulation not only verifies internalization, but it can also, in addition, be continued to the dissemination level of the taxonomy. The teacher must observe the simulation carefully to be able to verify internalization of planned behavioral objectives. Simulation serves to verify internalization in that simulation cannot be accurate unless the learner has internalized the necessary skills and behaviors. The teacher must further seek to help the learner verify within himself the accomplishment of these new behaviors. Learners usually enjoy simulation in that, with this strategy, they can show internalized behavior without ego threat. They can make mistakes or differentiate behavior in a role outside the self and thus can fully recognize the internalized behavior and can receive public support for it. Learners want to discover how new behaviors are going to affect them. Simulation is an excellent strategy to help them find out.

COMPARATIVE–CONTRASTIVE ANALYSIS

The learner applies internalized skills to new situations. This strategy involves analysis of similarities and of differences in systems or situations, and the teacher can use the interrogative mode. When new behaviors have been manifested, it is important to have the learner investigate these behaviors in a challenging setting. Comparative–contrastive analysis does this. It is, in a sense, the application to the internalization level of the research strategies and skills first accomplished at the identification level. With this strategy, the learner really deals with the learning experience in the inquiry mode. The learner encounters new forms of expression and must demonstrate mastery in different terms and in different contexts. Involved in this strategy are the skills of critical thinking and problem solving (see Chapter 7). By now the reader is aware that there is a taxonomic sequence for both problem solving and critical thinking and that the teacher must prepare the learners in these so they can be ready to use those skills in this strategy. Thus, the teacher works through the process with the learner by challenging and remotivating the learner as needs arise and as accomplishments are made. The teacher here must see that learners use and do something to test or challenge the internalized behavior. This strategy moves the process of internalization beyond skill reinforcement, re-creation, and role playing to a challenge-level usage of the newly internalized behavior. This strategy produces, beyond newly learned skills, a level of intrinsic internalization not necessarily apparent with prior internalization strategies. The learner is required, in a sense, to justify the efficacy of the new learning through analysis and through application to new contexts. It provides the opportunity for the learner to begin to multidimensionalize new skills. Although the dominant role of the teacher is that of a sustainer, the teacher must provide the challenge of new situations in which to test out the behavior. Few behaviors are fully internalized until they have met the test of a full-fledged challenge. This strategy offers such a challenge.

SUMMARIZATION

Here, a review of the experience takes place. The learner identifies with the totality of the experienced activities. The learner develops reports and tests out skills. Before completely internalizing a skill as a long-lasting personal behavior, the learner must have ordered and summarized the skill internally and within the social context. This strategy is appropriate to that need. The learner looks at and reviews the experience and begins to see the totality of it from exposure to internalization. The learner notes the process and the problems encountered along the way. The individual learner should be able to state or in some way to show new internalization of behavior. Summarization should be accomplished in some form before dissemination can become the final step in the taxonomic learning sequence. It can be accomplished through a written or through an oral summarization, but whatever the format chosen, this strategy becomes the final step before dissemination. Without incorporating summarization, an experience can be fragmented and may never come to full fruition in dissemination. Learner–teacher and learner–learner interaction is often a pertinent strategy at this point. Role playing or simulation can be used as a basis for the summarization process as can a written, learner-prepared report. The teacher needs to be sensitive to the need for summarization and to be ready to provide for this need by availing to the learner an appropriate format and by making certain that the learner has included in the summary those essential elements that both demonstrate the process and can provide a basis for making choices concerning dissemination. The teacher, by the very nature of this strategy, must have kept careful and complete records for the learner wherever needed. Once this has all been accomplished, the learner is ready to elect for dissemination and the teacher's task is to provide the learner with these opportunities.

## DISSEMINATION

### REPORTING

The learner prepares a written or graphic report sharing, showing, or explaining an experience. This is the written or graphic level of dissemination. The learner now presents, posts, displays, or makes available for review a summary statement, report, or graphic representation of a learned experience. Dissemination, by its very nature, is the learner's voluntary expression. No one, not even the teacher, should force or require dissemination. It can and should be encouraged and opportunities provided for it. Reporting that is required falls into the comparative–contrastive analysis (4.4) or summarization (4.5) categories. When it becomes voluntary and is intended to influence others in some way, it then moves to the dissemination level. One critical question a teacher can ask, however, is "May we display your report?" The learner has the right to decline and the teacher has the obligation to respect that right. If, however, the teacher has been careful to follow the role models through instruction, the stage has probably been set and in most cases there is no problem in obtaining dissemination activities from students. If the practice is encouraged, it will accrue as the teaching year progresses. A second factor to be considered is the teacher role of critiquor at this level. In this role a teacher's responsibility is to critique the dissemination process in order that the learner will proceed to new learning experience. Critiquing is, for the teacher, a deliberate effort. The teacher must see that the learners recognize the widest possible contexts of the experience so it can become, wherever possible, the springboard to a new learning experience beginning at the exposure level. Since learning never ceases, the teacher must see that this cycle is completed. Reporting is, therefore, an initial teaching strategy at the dissemination level of experience.

ORAL PRESENTATION

The learner orally shares, shows, or teaches the skills, the product, or the result of an experience. The sharing and the showing are voluntary. The activity can be a display if it is explained orally. In this strategy the learner formally presents in an oral form the internalized data, the modified behaviors, and the creative accomplishment of the experience. Although the teacher may have required a presentation previously as a summary, at the dissemination stage the report or work is presented voluntarily to enable others to share the experience. It also can become an imperative self-initiated dissemination effort. Oral presentation–reporting involves both levels of the dissemination category of the taxonomy. It may be either informational or homiletic. It must be remembered that an experience most frequently results in behavior modification at a performance level in which the dissemination is informational. When the experience leads to behavior modifications with a strong value orientation, the dissemination can become homiletic. An experience often can be disseminated at both the information and the homiletic levels. Teachers, in their taxonomic role of critiquors, must prepare for the dissemination by careful planning and by attending to the logistical problems involved in the strategy. They must also be the critics–evaluators who help build bridges between the current experience and a whole spectrum of possible ones. When learners display an inclination or a desire to disseminate an experience, this is a reward for the teacher and means the fruition of careful planning and the anticipation of new structures of experience. This, then, is a major teaching mode at dissemination and can be carried out in many ways ranging from oral presentations that illustrate the superiority of one style, process, or view over another, to projects designed to influence, to sell, or to advertise. Each learner needs the opportunity to choose this as a way of disseminating reactions to and learnings within an experiential learning situation.

## DRAMATIZATION

The learner makes a formalized presentation of internalized roles. This teaching strategy provides an effective vehicle whereby the student can express the experience. When involved in dramatization, learners are presenting for others to view those skills and internalized behaviors that they have accomplished through the experience. They are formalizing what they have learned for others to see. Dramatization can range from skits and pantomimes to a formal play. It can be written, produced, and played by the learners; it can be a professional play acted out by learners; or it can be a combination of both. Creative writing and preparation for the dramatization usually take place with the strategies used at the internalization level. Its function differs from that of role playing in that it is for public viewing rather than for the immediate set of learners. Furthermore, it is something the learners themselves choose to do. The role of the teacher again is to coordinate and to critique the dramatization. Within this role, the teacher must identify limitations and project new sets and behaviors while reviewing and evaluating the performance.

Dramatization has two thrusts. First, it provides a format for learner dissemination that can be either informational or homiletic, or both. Second, it offers an opportunity for the teacher to bring about new sets of possible experiences. The role becomes that of a critic and, in some contexts, even that of a manipulator. The teacher must make a bridge from the dissemination experience to an exposure to fresh experiences, so that the learner can begin again from the exposure level of the taxonomy. This can be a time of triumph and opportunity. There is triumph in the sense that, when well planned, the manifested behaviors are close to those foreseen at the exposure level of the experience. There is opportunity in the sense that the teacher now has a fresh responsibility to renew the experience with another set of motivators and stimuli. At the dissemination level, the experience continuum and the teacher role come full circle.

GROUP DYNAMICS

This takes place in the interactive context of dissemination and utilizes the social dimension of experiential dissemination. Dissemination is a uniquely extrinsic and social dimension of the experiential activities of learners. By its very definition, dissemination is social in context and requires individual interaction with one or more other people. Group dynamics perhaps best summarizes this social context. The basic requirement is social interaction, which may take the form of telling, sharing, discussing, questioning, recruiting, selling, reporting, answering, debating, speaking, and any other element of personal and group dynamics. Of course, participatory levels of group dynamics and expressed skills in interaction develop as learners mature. Yet group dynamics is as alive and appropriate a strategy for the prekindergartener as it is for the mature adult learner. The major decision the learner makes is to become involved in the group process in order to disseminate an experience. This, of course, by definition must be a learner choice. The teacher is obligated to provide opportunities for this dissemination, however, because internalization that lies fallow can result in random weeds rather than a cultivated crop of rich new experiences. In the fire of dynamic interaction, the teacher in an evaluative–critic role can cultivate new experiences and can prepare the learner for a new harvest of ideas. Group dynamics, like all the other dissemination strategies, is enjoyable for the learner and is a real opportunity for the teacher. The teacher needs to realize that the rationale for dissemination is growth to new experience. When this rationale is met, growth is indeed possible.

SEMINAR

At the seminar there is total participant interaction in sharing ideas on the effect of the experience and on the beginning of new experiences. A seminar is a reasoned coming together to discuss, to investigate, and to relate ideas about a particular subject or experience. It usually takes place in a relatively small group, seldom larger than classroom size. Those who participate in seminars are usually motivated by a desire either to influence or to test their present learnings. There is a common interest in the subject or the experience as well as a common interest in the interchanges within the group. In a seminar setting, it is expected that all participants will share views and ideas and will react to the development of the sharing and the interaction. The setting can often become a hotbed of new ideas and the teacher, in working with the learners to plan the seminar, needs to be aware of its potential for beginning new experiences. The teacher should be involved as critiquor in the seminar's progress and development. The teacher, as with other teaching strategies for dissemination, may need to take care of the logistics of the seminar so that the setting can promote fruitfulness. Like others, this strategy can be formal or informal, depending on the situation. Its development needs to be keyed to the maturation level of the learners. When the seminar is successfully concluded, the outcomes should include a range of possible new experiences for the learner and new sets of plans for the teacher. The direct purpose again is to start new experience built upon the recently completed activities. Seminars should not become a dissemination teaching strategy unless the objective is to move a completed experience to a new set of learning stimuli and to begin anew the taxonomic sequence of experiential learning.

## SUMMARY

These 25 teaching strategies summarize the interactive elements of the teaching–learning act. The teacher can recognize them as a lesson progresses and can observe or hear them on a videotape or an audio recording of a lesson. With these basic strategy classes that subsume within them many other related techniques and strategies, the teacher can begin the process of evaluating the teaching–learning act. This process of evaluation is discussed in the next chapter.

# 9 | Evaluating the Teaching–Learning Act

What is a good learning experience? Educators, parents, and students have studied that question for generations. Multiple responses have been put forward and some insights have been made. In particular, much progress has been made in terms of the way people learn. Yet, with all the data, research, and responses to the question, it still remains a cogent concern for educators, students, and lay people alike. Although we propose a response to the question, we do not presume to finalize an answer. We propose, rather, a new approach and a fresh insight to the teaching–learning act that may serve as a vehicle for further research and study of the problem.

There are the three basic elements to a good lesson that need to be noted initially. First, a good lesson requires careful planning. Admittedly, there is incidental and accidental learning, but in the context we are addressing are teacher-prepared learning sequences.

Essential to any teacher-prepared learning sequence is specific and sequential planning. Objectives should be visible and achievable; activities noted; resources gathered; teaching strategies defined; learner behaviors anticipated; problem solving, creativity, and critical thinking accounted for and built into the learning sequence; and assessment procedures projected. Second, a good lesson requires appropriate execution. The teacher needs to be alert to the necessary changes in role behavior as the lesson progresses. There is the role change intrinsic to taxonomic progression that was discussed earlier: moving from motivator to catalyst, then to moderator, to sustainer, and finally to critiquor. Through a learning sequence, the teacher must be sensitive to the individual needs of learners. Third, a good lesson requires evaluation of the students as well as teacher evaluation of resources, of strategies, and of personal performance. It is relatively easy to evaluate and assess student learning and it is expected that student evaluation is integral to the lesson sequence. Evidence of learner achievement is helpful in pointing out whether learner achievement has increased or that new skills have been learned. The question of why the learner achieved is not addressed in this type of product evaluation. Product evaluation is the keystone of teacher planning. Summative evaluation or, in another context, learner achievement itself, is a necessary criterion in planning for learner achievement. Teachers must organize around product evaluation for planning purposes in order to develop a teaching–learning process that can more effectively help students learn. Most objective studies of schools and of education have dealt with the product or summative effect of the school's work and have made judgments as to the quality of education, based on product evaluation. Most taxonomies have been sequenced around product evaluation.

Process evaluation, on the other hand, deals with what the teacher does that leads to product achievement. It looks at the teaching–learning act itself as an entity with the basic questions being why learner achieved and what methods or strategies were most effective in gaining that achievement. The experiential taxonomy is unique among taxonomies in that it provides an organizational format for process evaluation and a means of professionally analyzing the teaching–learning process. It is this need for determining the nature of a "good" lesson that can substantially affect the quality of the product. If teachers have a model for process evaluation and can effectively use it, the quality of the product will be enhanced. It is in this sense that we stress that a vital part of the third element of a good lesson is teacher self-evaluation. Teacher self-evaluation looks at the process, attempts to isolate and identify those elements that fostered or that deterred learner growth, and then uses that self-evaluation to improve the quality of succeeding teaching–learning acts.

Our response to the problem of improving the quality of the teaching–learning act is to propose a taxonomic model for teacher self–evaluation that is functional and easy for teachers to learn and use. This model was tested in a research study. Teachers have evidenced ability to master this model in workshops ranging in length from 4 to 8 hours. Planned 2-day workshops can effectively prepare teachers to use this model. This teacher self-evaluation model includes the following elements:

1.0 The experiential taxonomy
2.0 The teaching strategies and their descriptors
3.0 Teaching role models
4.0 A coding system
5.0 Interpretation and analysis of the coding
6.0 Replanning procedures

## THE EXPERIENTIAL TAXONOMY

The experiential taxonomy has been explained earlier. What is important for self-evaluation is, however, that the teacher be aware of more than just the five major categories.

The teacher should be able to relate them to the teaching–learning act. To self-evaluate taxonomically, the teacher should be able to link functionally the categories of the taxonomy to the major components of the teaching–learning act. The teacher will need to recognize at what points in a lesson sequence the activities and attendant strategies move from one level of the taxonomy to another. In this self-evaluation process the teacher must recognize class interaction and/or activities demonstrating exposure, participation, identification, internalization, and dissemination. The teacher, therefore, needs to know the categories of the taxonomy with some degree of internalization and to recognize progression through the taxonomy on an audiotape. It should be noted that an audiotape of a lesson can be made personally by the teacher and is preferred in the process of self-evaluation. Videotapes are also appropriate but are technically more difficult to manage, are more costly, and often require other personnel in the room. The audiotape can be operated easily by the teacher and reflects those elements of the teaching–learning act necessary for self-evaluation.

## THE TEACHING STRATEGIES AND THEIR DESCRIPTORS

Key to teacher self-evaluation is an understanding of the teaching strategies. When a teacher is listening to an audiotape or viewing a videotape of a lesson, it is important to be able to identify the teaching strategies used in order to follow and to analyze the process of the teaching–learning act. This recognition of strategies and how they are used will greatly enhance teacher self-evaluation. The teacher should spend time becoming familiar with the chart on pages 108–111, which lists the strategies and the brief descriptors of those strategies. It is important also that the teacher be aware of the more complete definitions of strategies given in the previous chapter. Wherever possible, these should be available for teacher review before coding begins. It is initially necessary for the teacher to have, for reference purposes, the chart previously noted for quick referral as the review and analysis of the teaching–learning act progresses. A good way to learn the strategies and the functional use of the chart is to spend some time in an in-service or group setting where teachers have the opportunity to hear tapes, identify strategies, and share interpretations. For a suggested organization of such a teacher in-service, see Chapter 11. Teachers should audiotape their own lessons in order to build a tape library for using the process of self-evaluation. One need not commit the descriptors to memory, but thorough familiarity is needed. Continued use of the previously mentioned chart through the coding process is helpful. It is important also to learn the number key for teaching strategies for use in coding. This numbering key to teaching strategies is shown on p. 146.

## TEACHER ROLE MODELS

It is apparent that a teacher's role changes as a learning sequence progresses taxonomically. Teachers should be familiar enough with these roles that they can recognize how their role changes both subtly and apparently as a lesson moves in increments through the taxonomy. One can hear on an audiotape and visually see the role of the teacher change as a learning sequence moves through the taxonomy. The five teaching role models (see Chapter 6) are keyed directly to the taxonomic levels. These teaching role models are motivator at exposure (1.0), catalyst at participation (2.0), moderator at identification (3.0), sustainer at internalization (4.0), and critiquor at dissemination (5.0). Although these role models are not integral to the coding system, they do serve to alert the teacher to changes in teaching strategies and to movement along the taxonomic sequence and do provide another dimension for teacher self-evaluation.


## A CODING SYSTEM

Previous coding systems for teacher evaluation have often been either too cumbersome for teacher use, too complex for functional teacher use, or not designed for teacher self-evaluation. Coding systems for teacher self-evaluation should be understandable, straightforward, and easy to use. The Experiential Taxonomy Teaching Strategies Coding System presented here meets these three criteria. The system can be learned by teachers in a relatively short time given some knowledge of the major categories of the taxonomy, awareness of the taxonomic teaching strategies, and some perception into teaching role models. Workshops ranging from 1 to 2 days have sufficed to teach professionals how to functionally code for self-evaluation. The process can be shortened for those with prior knowledge of the taxonomy. With an understanding of the taxonomy and the teaching strategies descriptor chart available for reference, the teacher can begin the process of coding.

Coding is done on a sheet that is set up to code 20 minutes of a lesson. Minutes run down the left side of the page (see sample, Figure 9.1). If the lesson lasts more than 20 minutes, additional sheets should be used. The number of the teaching strategies 1.1, 1.2, 1.3, and so forth through 5.5, shown on the chart in the previous chapter, are listed across the top of the page. The first number refers to the taxonomic level and the second to the specific teaching strategy of which five are identified at each taxonomic level. For example, 2.5 refers to ordering and 4.1 to skill reinforcement. There is a square in

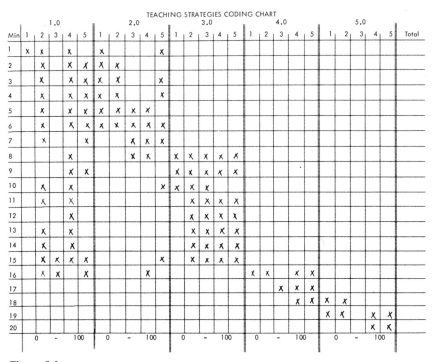

**Figure 9.1**
An example of a coded lesson.

each minute for a check to be placed if the strategy is used. Once the coding sheet and the numbering system are understood, the next step in the coding process is listening to the tape and actually coding the teaching–learning process. The coding itself is done on a minute-by-minute basis. The teacher places a check in the box for each strategy noted during a particular minute. In the first minute of a lesson, for example, a teacher may hear incentive conditioning (1.1), data presentation (1.2), and directed observation (1.4). A check would be placed in each square representing those strategies in Minute 1. In Minute 2, the teacher may be involved in more data presentation (1.2), in some data exploration (1.5), and in some student modeling–recall (2.1). A check is placed in those squares for Minute 2. Minute 3 may have a demonstration (1.3), data exploration (.15), modeling–recall (2.1), and expanding data bases (2.2). If so, the appropriate squares are marked. And so the system goes. Coding becomes easier the more frequently it is used. Not all categories of the taxonomy need to be present in any one lesson sequence and there are reach-back strategies (see pages 149–150) that are apparent in many lessons. Teachers should also remember that individual interpretations may vary in terms of some strategies. For example data presentation, directed observation, data exploration, and modeling–recall can be differently interpreted. More important, however, than the existence of varying interpretations of specific processes and strategies is the fact that general patterns emerge as the lesson moves in increments through the taxonomy and that the analysis and interpretation of these general patterns are the important key to teacher self-evaluation. Group coding, as noted earlier, can be extremely helpful in early stages of coding to gain confidence and familiarity with the process and to lessen varying interpretations. The teaching strategies are as follows:

1.0 Exposure
    1.1 Incentive conditioning
    1.2 Data presentation
    1.3 Demonstration
    1.4 Directed observation
    1.5 Data exploration
2.0 Participation
    2.1 Modeling–recall
    2.2 Expanding data bases
    2.3 Dramatic play
    2.4 Manipulative and tactile activities
    2.5 Ordering
3.0 Identification
    3.1 Field activities
    3.2 Using data
    3.3 Discussion–conferencing interaction
    3.4 Hypothesizing
    3.5 Testing
4.0 Internalization
    4.1 Skill reinforcement
    4.2 Re-creation
    4.3 Role play–simulation
    4.4 Comparative–contrastive analysis
    4.5 Summarization
5.0 Dissemination

5.1  Reporting
5.2  Oral presentation
5.3  Dramatization
5.4  Group dynamics
5.5  Seminar

## INTERPRETATION AND ANALYSIS OF THE CODING

Coding, once it is learned by a teacher, can be enjoyable and even an intellectual challenge. Yet it has little or no professional meaning unless the teacher is willing to spend time objectively analyzing and interpreting the teaching–learning act. One of the first things that the teacher should do is to note the total number of checks in each column of the coding sheet, thus learning how often a particular strategy was used in the lesson. A teacher can enumerate the number of times, for example, that the learners explored data (1.5), how many times the learners were involved in manipulative and tactile activities (2.4), how often the learners were hypothesizing (3.4), what was done in skill reinforcement (4.1), and how frequently the learners disseminated what they had learned. Following the column totals, the teacher should find the totals at each level of the taxonomy. This can give an overall pattern to the teacher of what happened in the lesson. The teacher now has a quantitative lesson profile. This quantitative dimension may also be graphed or charted to show more clearly what teaching strategies were used and how far into the taxonomy the lesson progressed. More importantly, the teacher needs to know the approximate amount of time spent at each taxonomic level and which levels of the taxonomy the students experienced or achieved at the end of the lesson. Of equal concern to the teacher is whether the learners achieved the level of the taxonomy that the teacher had planned for them. A review of the lesson profile and an interpretation from the quantitative enumeration of teaching strategies lead the teacher to the qualitative aspect of lesson analysis.

Basically, this qualitative aspect deals with the question, What made this a good (or limited) learning experience? In a qualitative interpretation of lesson sequence, one needs to assess the lesson from the perspective of certain generalizations about the teaching–learning act that are basic to the Experiential Taxonomy Teaching Strategies Coding System. They were developed as a result of extensive coding and analyses of lessons with teachers during research on the coding system. The following generalizations should be considered during any lesson sequence analysis on the qualitative level.

## MINIMUM TAXONOMIC LEVEL FOR LESSON SUCCESS

Any successful lesson must achieve a 2.5 taxonomic level (Steinaker & Harrison, 1977). Even at the 2.5 level the teacher is dependent on past student experience, on concurrent student experience, or on a future experience, for the objective to be internalized. The teacher can plan guides leading through the taxonomy based on lessons ending at the 2.5 level. If the teacher stops here without further learning sequences, the learner must bear the responsibility for further progress in the experience.

## TEACHER–LEARNER ROLE DOMINANCE

The teacher, the learner, or both can have role dominance as the lesson progresses through the taxonomy. In reviewing the teacher role models (Chapter 6) one notes how the role progresses from teacher dominance to learner dominance. These differentiations in roles can be heard on the tapes.

## TAXONOMIC GOALS

A unit (two or more linked lessons) may have long-range taxonomic objectives and daily lesson objectives. With several skills involved, the lesson may be at the exposure level in terms of long-range objectives, at the identification level in terms of short-range objectives, and at the internalization level in terms of a particular lesson objective. Thus, one must view each lesson in terms of its relation to the total learning sequence and to the activities around it. It is also possible to reach internalization on a particular lesson that is the exposure level of the larger learning context.

## REACH BACK

Reach back occurs when the teacher, for review or summary purposes, needs to use earlier taxonomic teaching strategies to reaffirm learner readiness for progression along the taxonomic sequence or to summarize the experience. Strategies 1.2, 1.3, 1.4, and 2.5, as well as others, have merit as reach-back strategies to review data, provide supplemental data, focus, or order thinking so that learners may move to higher levels of the taxonomy.

*DATA GENERATION*

Teachers can sense when they have generated enough data at exposure (1.0) and participation (2.0). A teacher must not spend too much time at exposure or participation. Students reach their data tolerance at varying rates depending on mental age and maturity, and a teacher must be sensitive to this. The teacher should provide enough data and have the learners explore it, interact with it, augment it, and move on when they reach the level of data tolerance. If the teacher feels that learners need a great deal of data, the data should be presented in two or more sessions with each session ending with ordering (2.5), thus allowing the learners to deal with the data at their level of data tolerance.

*TIMING*

Timing is an integral element in the teaching–learning act. Teachers can sense when their timing is off. Learners tend to lose interest and teachers must reach back more often to reorient the lesson to an appropriate closure. If the teaching–learning act indicates reach ahead or reach back, the teacher must be flexible enough to do this.

*TRANSITION TO DIFFERENT TAXONOMIC LEVELS*

Ordering (2.5) can be used as a transition even when learners are at the identification (3.0) or the internalization (4.0) levels of the taxonomy. It is used to reorder the activities toward accomplishment of the lesson objectives. In addition, it can be used for organizational purposes. Ordering is a strategy that can reaffirm, refocus, and provide direction. It is an essential strategy for successful teaching.

## EXTENSION OF LEVELS DISCUSSION

After students have had a field activity (3.1), teachers can extend identification (3.0) and internalization (4.0) levels of discussion without relying on a 2.5 reordering for much longer periods of time. Discussion here has meaning and perspective; the students have identified with the experience and at these levels student input is much greater and has more specific relationship to the learning objectives and to planned outcomes.

## FIELD ACTIVITIES

A critical strategy for promoting incremental progress through the taxonomy is field activities (3.1). Without field activities or the "doing" strategy, there may be limited progress through taxonomic learning. With creative planning, effectuating this strategy can bring about real learner identification with the learning experience.

## TEACHING STRATEGY CLUSTERS

Field activities (3.1), using data (3.2), and discussion–conferencing interaction (3.3) are linked and generally go on concurrently. A teacher trained in coding can identify these strategy clusters. As clustered strategies, they tend to ensure learner identification with the learning experience.

## RESEARCH ORIENTATION

The whole series of teaching strategies at the 3.0 level are linked together in a research orientation that takes learners through the scientific method. It is at this experiential level that the higher levels of cognition begin to be apparent and learners begin to gain real commitment to the learning experience.

## INTERACTION AT THE INTERNALIZATION LEVEL

Learners and teachers relax when they know that they have achieved planned objectives through the internalization level. The voices and actions of both teacher and learners change when the internalization level is reached and the teacher role as sustainer becomes apparent.

*THREE TAXONOMIC LEVELS IN ONE CODING*
*TIME FRAME*

When three or more taxonomic levels of teaching strategies are apparent in the 1-min coding time frame, the learning experience is greatly enriched. This mix of strategies means that the teacher is involving the students in multiple activities leading to the internalization and the dissemination. At this time the teacher knows students have learned and is using reach-back techniques to sustain and to augment the internalization.

*DISSEMINATION TEACHING STRATEGIES*

Dissemination teaching strategies need to be as carefully planned as any step in the teaching–learning act even though dissemination is a voluntary learner activity. The role of the teacher as critiquor must be carefully and specifically planned to move the experience to other contexts.

The preceding generalizations about lessons and about their process development can be the basis on which a teacher may analyze and interpret a lesson. A teacher who is coding needs to be aware of these generalizations and of other aspects of lesson interpretation. It should be remembered that any interpretation of a lesson is based on the prepared product objective. Process evaluation is based on whether the process of achievement succeeded and why and how it did or did not succeed. Both product and process evaluation are necessary for any continuing planning. A teacher should also consider that successful learning experiences require the learner to achieve the identification (3.0), the internalization (4.0), or the dissemination (5.0) levels of the taxonomy. If most lessons terminate at ordering (2.5) or early in the identification (3.0) level of the taxonomy, then the teacher should carefully examine the lesson planning.

## REPLANNING PROCEDURES

No analysis or interpretation has real value unless it has some effect on teacher planning and on change in future teaching–learning sequences. As teacher self-evaluation occurs, a concomitant growth in planning and in general professionalism often occurs. As teachers begin to take into consideration what has happened in the classroom as they have evaluated their own teaching through the Experiential Taxonomy Teaching Strategies Coding System, they begin to realize that they now possess a tool for effective planning. When they plan after the self-evaluation, teachers are better prepared to write objectives more clearly, more specifically, and in behavioral terms. They are also able to define more clearly the activities relating to the objectives and able to sequence them taxonomically. They know more about themselves and about their own teaching styles. They know which strategies work for them or for their learners and they have a real concept of the teaching–learning act that can only augment their professional skill. With these activities sequenced appropriately, the teacher is then in a position to select taxonomic teaching strategies that will enhance student learning. The teacher can then implement the lesson, tape it, code it, and evaluate it and, thereby, have a personal taxonomic process evaluation tool that can result in greater learner mastery of the identified objectives. The teacher can easily learn the coding process, as mentioned earlier, in a workshop of no more than 2 6-hour days and sometimes even more quickly. Furthermore, the more frequently a teacher self-evaluates, the more effective the teacher becomes in enhancing student achievement. In our experience, this type of self-evaluation during a unit or a lesson progression has contributed to learner achievement and also results in professional development and in increased teaching skills.

# 10 | Teacher Change and the Experiential Taxonomy

**P**rofessional development is one of the necessary elements in any teacher's career. The beginning point in this development is the first classroom assignment and from there the process of professional development continues. The usual process of teacher change is a progression of extrinsic development through professional courses, workshops, and inservices, with the teacher then applying the learned skills to the classroom. These elements of professional development are essential, but there is also another dimension to teacher change. This dimension begins with the teacher, with the teaching–learning act, and with the teacher looking at personal performance in the classroom. In other words, teacher change and professional development can begin as an intrinsic part of the daily task of instruction. Such self-inspection can become the impetus for teacher change and for professional development and can result in improved student learning. As an intrinsic part of the teacher's daily interaction with students, this process of change can be enhanced through teacher use of the experiential taxonomy as the process organizer for change. Like teaching strategies and like the sequence of learning activities, teacher development can be defined and organized taxonomically. Following is a taxonomic framework for teacher change:

- EXPOSURE: Understanding the functional use of the taxonomy as a basis for professional growth
- PARTICIPATION: Using the taxonomy for planning, for implementing, and for evaluating student learning
- IDENTIFICATION: Doing process evaluation through coding, analyzing, and interpreting classroom interactions in the teaching–learning act
- INTERNALIZATION: Manifesting changed professional behavior in terms of planning, implementing, and evaluating the teaching–learning act
- DISSEMINATION: Teaching others the process of professional growth

This format, in succinct form, is the taxonomic sequence for intrinsic teacher professional development and for teacher change. It is a simple process involving carefully planned sequences of activities with emphasis on the teacher as self-evaluator. The resource leader charged with the responsibility for teacher development can use these taxonomic steps as a basis for planning. It will be noted that there is an overall *process* in this format as the teacher progresses through the taxonomy and that there is, likewise, a *product* or skill developed at each taxonomic level. The teacher will profit professionally from accrued skills at each step in the process no matter when or under what circumstances the experience may end. This sequence, like any learning experience, can have an impact on teacher change at whatever level of the taxonomy the teachers involved in the process do achieve. To fully understand this process and the product or skills that the teachers involved achieve, it is necessary to examine it in terms of each level of the experiential taxonomy and the elements of teacher change implicit at each of the major taxonomic categories.

## EXPOSURE

This step in the process of intrinsic teacher development focuses on teacher interaction with the taxonomy itself. Here the teacher in a workshop, in a professional course, at in-service training, or even through professional reading becomes familiar with the taxonomy and with its use in planning, in implementing, and in evaluation. The teacher, at this step, gains enough information and knowledge about the taxonomy to elect to continue personal professional development if the in-service does not cease at this point. This is really the motivational step with the leader providing the motivation or the teacher being personally motivated to get information about the taxonomy not just for the data, but for their value as a basis for professional growth, with the Experiential Taxonomy Teaching Strategies Coding System. By the close of in-service at this level, the teacher should have read, discussed, and related the taxonomy to the classroom function.

The teacher involved should further recognize the taxonomy as possibly functional for classroom planning, teaching, and evaluation through functional skill in coding lessons with the Experiential Taxonomy Teaching Strategies Coding System. Because many teachers are not generally self-motivating, someone else usually plans this beginning step toward greater professionalization, generally the resource teacher or instructor. This is a step in teacher change that may be somewhat intrinsic; but, as the participant begins to explore the data, certainly the process of professional development becomes progressively more intrinsic. The product—the coding skills that the teacher acquires—in itself indicates a professional growth step and can inspire self-motivation. The major task of any in-service leader or instructor, however, is not simply to teach a skill but rather to motivate participants to continue the in-service and to show an understanding of the taxonomy. Evidences of an understanding of the taxonomy in relation to the teaching–learning act and to professional growth come as the teacher begins to show an ability to code lessons using the taxonomy. The steps leading to this ability to code are usually somewhat as follows:

- awareness of the taxonomy through reading or through presentation
- discussion or thinking about the taxonomy in order to understand its elements
- focusing on curriculum development as a taxonomic process
- developing the ability to code lessons using the Experiential Taxonomy Teaching Strategies Coding System
- understanding how coding works as a means for self-evaluation
- coding a lesson successfully either from one's own tape or from the tape of another teaching–learning act

Once the process of coding is mastered, the participant has completed the exposure step of the process. The time involved will, of course, vary from participant to participant. Upon deciding to apply the new understandings to the curriculum, the teacher advances to the participation level of taxonomic change, and another taxomonic level of teacher change becomes pertinent.


## PARTICIPATION

The teacher now begins to translate understanding and acquired skills into doing and using and has thus learned two processes. The teacher, first, is aware of the process of taxonomic curriculum development and, second, knows how to code, analyze, and interpret a taxonomic lesson. In the participation stage, the teacher will express that knowledge in the actual preparation of a taxonomic curriculum unit, will implement the developed unit, and will self-evaluate individual lessons within the unit by coding them. Because there are several distinct steps in the total experience design for curriculum development, it is usually wise for the teacher to develop initial teaching units with strong support from an in-service leader and/or from peers involved in the same professional development process. Here the teacher needs to ask questions, to seek information, and to receive support as the in-service process continues. Someone must fulfill the felt needs of the teachers. In situations where the individual teacher is preparing a taxonomic unit for the first time, it is most helpful to do it in a small group setting where others are also developing units and where the resource leader can assist. In this way, the teacher can make greater progress toward the completion of an effective unit. Likewise, some modus operandi needs to be arranged during unit implementation so that the teacher can receive similar support during the instructional act. Support must not be seen as supervision but rather as the work of one interested practitioner with another. In evaluation, the same applies. We caution that, although product evaluation in unit implementation is very important, supportive evaluation at the participation level of taxonomic teacher change will tend more to inspect the processes of the instruction and the analysis of the teaching–learning act. The experiential taxonomy is the only extant taxonomy that really focuses on this kind of process evaluation. When the teacher begins to conceptualize process evaluation and to demonstrate the ability to prepare, to implement, and to evaluate a taxonomic teaching unit, the next step in the taxonomic sequence of teacher change becomes apparent.


## IDENTIFICATION

The teacher is aware of and has demonstrated the ability to use the coding procedures at an earlier taxonomic level in this process of professional development. While actually teaching a self-designed experiential unit, the teacher can begin the identification process. The teacher, during this identification stage, plans to tape an initial or early lesson in the unit sequence and then to tape lessons on a regular or a scheduled basis throughout the unit. These audiotapes, which are relatively easy to record (by simply turning a cassette tape recorder on "record" and leaving it operating during a class session), should then be used for coding with the Experiential Taxonomy Teaching Strategies Coding System (ETTSCS) described in Chapter 9. The teacher should code with the teaching strategies descriptor chart available as a reference. A review of the teaching strategies, preferably in a group setting, should precede the actual coding exercise. Although in many instances the teacher will do this initial coding individually, there should also be some arrangement for group coding. The purpose of group coding is to develop more skill and consistency in identifying teaching strategies and to provide for group analysis and interpretation of lessons. We have found through conducting workshops that group coding greatly enhances individual coding later. Sometimes individual analysis and interpretation may be necessary until shyness wears off, but group discussion, analyses, and interpretations are very healthy professional activities and they should be encouraged whenever possible. The product at this taxonomic level is the manifested skill in coding and the analysis of the teaching–learning act.

The in-service leader plays a key role here in providing data and positive assistance whenever needed, and in supporting and encouraging analysis and interpretation. Since the in-service leader is a vital link in professional development at this and at all other taxonomic levels, this person should be well trained and should have much experience in coding, analysis, and interpretation. It should be noted that the whole process of coding, analysis, and interpretation should be repeated and should continue throughout one's professional service until it is completely and mutually internalized and one can follow the process of the teaching–learning act without the necessity of coding. Teachers can always grow and develop in their skill and their ability to foster student learning by improving the teaching process. When the behavior changes occur, a new level in the taxonomy of teacher change becomes apparent and resultant teacher development can occur. This new level of change is, of course, the internalization level in this taxonomy of teacher change.


## INTERNALIZATION

When a teacher achieves this level in the taxonomy of teacher development, there is an overt manifestation of change in professional behavior in the teaching–learning act. The teacher will, for example, spend a much higher percentage of the teaching–learning time using strategies at Levels 3.0, 4.0, and 5.0 of the experiential taxonomy than before the process began. Observed teacher change not only can be seen but also can be validated by being coded and analyzed. Furthermore, teachers themselves will recognize the changed behavior and their students will have the opportunity for increased growth in the classroom of a professional who has truly internalized the experiential self-development approach to the teaching–learning act. Even at this stage, however, the process of professionalization is not complete. That comes with dissemination.

# DISSEMINATION

When teacher change and professional growth have been manifested, one needs to culminate the process through its dissemination. Education is not an esoteric craft but rather a profession in which mutual cooperation and professional interaction are the rule rather than the exception. When a teacher has truly internalized and manifested positive professional growth, it is incumbent upon the individual teacher to help others achieve new observable professional skills that can result in increased student learning. Again, an administrator, an in-service leader, or an instructor can be a key person in setting the tone for professional sharing, as well as for professional cooperation and interaction. This leader's role is now to provide the opportunity for teachers to express their experience voluntarily. The process might begin in a simple sharing situation—a teacher assisting another teacher—teachers could share at an in-service meeting or a workshop where the teacher can interact with peers around the process of teaching and working with learners and around personal experience with the taxonomy. As the opportunities for dissemination continue, more and more responsibility for presentation and leadership could accrue until the teacher assumes the full responsibility for planning and conducting an in-service workshop.

Two kinds of dissemination of in-service growth need to be mentioned. They are informal dissemination and formal dissemination. The informal in-service dissemination is usually presented in a familiar context where there is a kind of warm, simple, direct sharing and growing. Much of this kind of in-service takes place in a comfortable and positive atmosphere and dissemination is seen as the voluntary sharing of a mutual experience that has been internalized. Staff meetings, department meetings, and meetings of peers within one's own school are usually the settings for informal in-service. These in-service sessions need to be carefully planned with expected outcomes, but many of the constraints of formal dissemination can be avoided. Due to the nature of formal dissemination, with its frequent time limitations and expected outcomes, careful planning is necessary to make it a valuable and a rewarding experience. When the teacher attends both the informal and the formal in-services as a professional choice, the taxonomy for change in teachers culminates.

In this discussion of teacher change and the experiential taxonomy, a taxonomic sequence for achieving change has been presented in brief form. It should be noted that at each taxonomic level, growth can occur and that there is a professional integrity to the accomplishment of objectives at that taxonomic level. In view of this, one can see that change can be planned by administrators, by in-service leaders, and by teachers themselves through a series of activities keyed to the major categories of the experiential taxonomy. It has been found, for example, that at the exposure level most teachers need a minimum of 2–6 hours in a workshop setting to recognize and understand the functions of the taxonomy as a basis for professional growth. Some, particularly those who have no prior knowledge of the taxonomy, may need more time. Mastering the taxonomy for planning, implementing, and evaluation at the participation level takes longer because of the actual involvement in classroom teaching necessary. Close adherence to the steps in the Total Experience Design, in Chapter 4, will help a teacher in planning a curriculum. This crucial step needs careful guidance from someone experienced in the Total Experience Design for curriculum development, implementation, and evaluation.

Teacher development at the identification level can and should go on concurrently with the participation level activities. Coding and its analysis and interpretation fit well within the context of unit implementation and evaluation. Likewise, the teacher can note changing behavior as the unit progresses. This is often a real impetus for further growth since there need not be a great time lag between the analysis and the interpretation of a coded lesson and the changed teacher behavior. Growth can occur naturally and very quickly. Internalization comes when the teacher recognizes that changes have occurred and is able to demonstrate those changes through self-evaluation and through manifested learner growth. Dissemination generally begins in the informal mode and can subsequently, in many instances, move to the more formal setting of in-service and professional meetings.

The reader can see how the steps are linked together. They flow sequentially and easily from one level to another and, in sum, make the process of teacher change and professional growth a matter of natural professional activity. In the next chapter we will identify and spell out particular in-service sessions for training in this taxonomy for teacher change.

# 11 | Staff In-Service Programs

The previous chapter described the process through which the teacher progresses when the experiential taxonomy is used for professional development resulting in changed teacher behavior in the classroom. It is the intent of this chapter to provide those steps necessary, first, for conducting workshops for using the taxonomy, and second, for structuring workshops in other areas using the taxonomic levels. Thus, those workshop leaders who desire to use the taxonomy as a vehicle for teacher development will have a guide and those who wish to ensure that other in-service programs include all necessary steps will have a process.

## IN-SERVICE FOR USING THE TAXONOMY

The in-service leader responsible for the organization and the implementation of a professional development program must plan carefully not only for the total group but also for those individuals of varying needs and degrees of experience and differing ability to move through the taxonomy. It is the leader's responsibility to plan outcomes and the process for achieving those outcomes. At each level of the taxonomy for teacher change there are particular achievement objectives that the planners should identify specifically. The objectives should include the expectancies noted for each level (see Chapter 10) and those additional expectancies pertinent for particular groups of teachers and/or paraprofessionals. With the objectives carefully in mind, the next step is to identify the process by which the objectives are to be achieved. The planners should give careful attention to this activity because it is through a planned process that outcome expectancies are fulfilled. It should be noted that the presenters and the participants are to evaluate both the process and the outcomes in terms of achievement and of effective process. A possible in-service program at each taxonomic level is outlined in the following. We will suggest alternative programs that should achieve similar goals. The five steps in the taxonomy for teacher change noted previously are

| | |
|---|---|
| EXPOSURE: | Understanding the functional use of the taxonomy as a basis for professional growth |
| PARTICIPATION: | Using the taxonomy for planning, implementing, and evaluating student learning |
| IDENTIFICATION: | Doing process evaluation through coding, analyzing, and interpreting classroom interaction in the teaching–learning act |
| INTERNALIZATION: | Manifesting changed professional behavior in planning, implementing, and evaluating the teaching–learning act |
| DISSEMINATION: | Teaching others the process of professional growth |

Presentations or professional programs for teacher growth can be built around objectives relating to each of these behaviors. In-service and professional programs can be designed to accomplish only one identified taxonomic goal (level) or they can be linked from one taxonomic level to another. Each can have integrity and achievable results in terms of teacher development. For example, a workshop or series of workshops could be designed at the exposure level simply to help teachers understand the functional use of the taxonomy. If leaders are expecting to move incrementally through the teaching strategies, one or more sessions should serve to achieve the objective for each taxonomic level. Planners and presenters should pace their presentations in such a way to allow the participants to interact with the ideas. All programs and workshops should adhere to the taxonomic structure so that the participants' perception of what they can achieve will be enhanced by the model that they are experiencing. Planners preparing in-service in taxonomic sequence should remember that in-service can be either informal or formal. They should adjust the format to the mode they use for the best results. In the informal mode there is more initial and consistent interaction because participants generally know and respect each other. In the formal mode presenters need to build conditions for participation into their planning and into the actual presentation. Interaction is essential to any in-service program for professional development. Specific in-service programs focusing on professional growth in terms of the taxonomy for teacher change will attend to the following categories.

## EXPOSURE

The central objective of a workshop or a professional program is demonstrated participant understanding of the functional use of the taxonomy as a basis for professional growth and as a basis for classroom teaching. As workshops are planned, this purpose must be known to each participant and must pervade each step. The planners must consider the professional level of each of the participants and must develop a plan that takes into account individual differences in professionalism and personality.

The initial workshop is considered most important because all future activities hinge upon its successful completion. The leader must continually recognize that this is the invitation to proceed further and that, although interest must be kept high, the participants' feelings of anxiety must be reduced. For this reason we have maximized attention to this workshop level even though those following are equal in importance.

Workshops designed to achieve the objectives of the exposure category would incorporate to five major levels to introduce and explain the taxonomy and to provide both the basis for and the invitation to further growth. The basic elements in the program are:

1.0 Introduction to the Experiential Taxonomy (Exposure)
    1.1 Need for a new taxonomy
    1.2 Its development
    1.3 The basic categories
    1.4 Definition of "experience"
    1.5 Taxonomic categories
2.0 Expanding the Taxonomy (Participation)
    2.1 An orientation walkthrough
    2.2 Implications for teachers, for learners, and for evaluators
    2.3 Mixed-up unit activity
3.0 Practical Experience (Identification)
    3.1 Curriculum development
    3.2 An introduction to coding
4.0 Expansion Exercises (Internalization)
5.0 Participants' Presentations (Dissemination)
    5.1 Informational presentation
    5.2 Design to influence
6.0 Review Summary Evaluation

Workshops designed to achieve the objectives at this exposure level would use the five major taxonomic categories to bring about an introductory understanding of the taxonomy and to provide both the basis for and the invitation to further professional growth.

Each of the elements has within it certain activities that relate to the accomplishment of the overall performance objective. Each element will also have a specific sequence in relation to participant expectancies. In looking at, in planning for, and in implementing these elements of the exposure level, the leader should think in terms of time involved. Although this depends on the participants, we have found this workshop to be accomplished best in two 5- or 6-hour periods. In planning each step, the following are useful.

*1.0 INTRODUCTION OF THE EXPERIENTIAL
TAXONOMY*

For background, the material in the Introduction of this book should be summarized and should be followed by presentation of the categories and the subcategories of the taxonomy itself. The participants should receive a copy of the taxonomy for personal use, and the leader should explain the major existing taxonomies and their specific contributions not only to curriculum development but also to teachers in their work with students. The leader should quickly explain that most taxonomies address only one aspect of human experience and that we need a unified, complete, and easy-to-use framework to guide development of daily lesson plans, of curricula, and of increased professionalism. The leader should cover the taxonomy's development as described in the early chapters and then present the five major categories with short explanations of each.

We have found it best at this point to present various dictionary definitions of an "experience," to present our own, and to discuss with participants their feelings as to what an experience is before going more deeply into the taxonomy.

A visual overview describing each category and then each subcategory is now most effective and can easily be presented through overhead projections or simple drawings and descriptions on large chart tablets. A time for questions and reactions is a necessary part of this introduction because it allows the leader to summarize.

## 2.0 EXPANDING THE TAXONOMY

Prior to discussing the implications of each level of the taxonomy, the leader briefly walks through the steps provided in Chapter 3 that could be followed in the learning of a new song. The leader should use this or any other example that appears common to participants to provide again a similar start-off point.

Two major activities follow that assist participants in making the bridge to reaching that point where they can begin to understand how the taxonomy can be an important asset to their profession. The first clarifies the implications the taxonomy has at each level for the teacher, for the student, and for the evaluator. After further expanding a category, the leader promotes clarifying discussions in a variety of ways. For example, when discussing the "exposure" level the leader introduces participants to the teacher's role (motivator), to the student's role (attender), and to the evaluator's role (observer). The leader asks questions such as: "What activities (or types of questions) do you see that can be used to gain student attention and desire to move ahead? What specific things would you look for to evaluate this?" We often divide up participants into groups of five when discussing similar questions regarding the "participation" level, have participants classify a list of possible activities and strategies when clarifying the "identification" category, and so on.

To both evaluate and summarize this part of the workshop, participants are provided a "mixed-up" unit and asked to try to put it together as they feel it should be if they were to follow the experiential taxonomy. Leaders should be cautioned to require only that the activities be placed in the proper order at this time. Upon completion, participants are presented the "school solution" and may challenge any parts with which they cannot agree.


3.0 PRACTICAL EXPERIENCE


At this point each of the participants should have available a copy of the Unit Development Chart (see Figure 4.3) for the curriculum development activity and of the Teaching Strategies and Descriptors" chart presented in Chapter 8 for the coding activity. The former chart lists under the suggested taxonomic level those activities, principles, and learner–teacher–evaluator strategies described through the text. The latter chart, to be used for coding, brings the strategies and their descriptors into a more cohesive format to assist in the professional development of the teaching act. Although exposure to curriculum development is presented first, it must be noted that the leader can easily reverse the sequence depending on the leader's own objectives. Each activity augments the other and moves the participant toward fuller understanding. Furthermore, the leader's awareness of individual needs and interests will assist in the determination.

Having completed the "mixed-up" unit activity, participants have been introduced to the format for unit development. Since this is still an introductory workshop, they are instructed that, in building a unit with the experiential taxonomy, participants are to rely upon their imagination alone and thus other tests and resource material are unnecessary at this time. A unit objective is provided or participants may wish to select their own to more properly reflect the level at which they teach. Since time must be allowed for sharing and for discussing units and to maintain a low level of anxiety, unit development is best done in groups. Whereas any awareness of taxonomic subcategories is useful in unit development, participants are instructed to concern themselves only with the five major categories at this time. When possible, copies of units developed should be reproduced for each participant prior to the discussion stages.

As they approach the coding activity, participants should review the Teaching Strategies and Descriptors chart, noting its more concise listing of teaching strategies and its applicability for classroom observation. The relationship between the taxonomy and the teaching strategies should be emphasized as participants become familiar with the descriptors and how they relate to classroom teaching. An interaction between participants as well as with the resource leader is profitable at this time. A number of questions will arise as participants begin to relate the descriptors to actual classroom learning. The leader may need to reassure some participants since no strategy is always easily identifiable and because some may fall into two or more categories. The important objective at this point is for participants to realize the functional use of the teaching strategies as part of the teaching–learning act.

Although the coding method presented here is relatively simple, for some it may take more time. Previously some coding systems with which teachers became familiar generated so much data and required so much recording that they became difficult to use and resulted in a negative reaction. Should this happen here, the resource leader's first task is to explain the coding form, relating it to the teaching strategies, and to walk the participants through the process. Resource leaders should have two audiotapes, both of which have been coded prior to the session. The form can be put on a transparency. Then, as the first tape plays, the participants mark blank coding sheets as the leader records on the transparency. The leader should stop after each of the first 5-minute segments and explain why certain categories were marked in specific ways. The first tape itself should be designed to illustrate all levels of the taxonomy and several teaching strategies. Parts of the tape that illustrate different levels of the taxonomy should be relatively obvious during this stage so that the participants can hear and code each level and the teaching strategies associated with it. After the group coding, when the teachers have compared the leader's coding with their own, leaders should review the Experiential Taxonomy Teaching Strategies Coding System thoroughly and discuss how its use can improve classroom teaching effectiveness. Participants are now ready for independent coding practice using the second previously coded tape. In this phase, each participant should have a copy of the chart of teaching strategies and descriptors as a point of reference. The second tape is then played while individuals code the lesson segment. When they finish, the participants discuss the lesson and gather and analyze a quantitative summary of the coding. The teacher then interprets the lesson by discussing those generalizations about the lessons described in Chapter 9. This latter procedure will enable the leader to obtain the necessary participants' reaction to their ability to code, analyze, and interpret a lesson.

## 4.0 EXPANSION EXERCISE

When they have completed the foregoing practical exercises, the participants will have gained, to varying degrees, an understanding of the taxonomy as a tool that, upon further training, they may use to increase their effectiveness. There is no certainty, however, that they will maintain this understanding even in the general sense without further activity. Expansion here involves extending the use of the taxonomy into other areas in order to remotivate and to rekindle interest in using the taxonomy in the participant's primary field. We have found that having participants select an area outside their immediate field is effective for this purpose. Participants select an area in which they feel the taxonomy may prove effective and outline how they might proceed through the five categories. Leaders are cautioned to make this an individual endeavor since this is an internalizing exercise, given only once, and thus maximum benefit has come from only one opportunity. They should suggest areas that the participants might use, such as selling, advertising, counseling, campaigning, public speaking, and studying.

## 5.0 PARTICIPANTS' PRESENTATIONS

Workshop leaders may wish to include two steps as the finale of their workshop. Both measure the effectiveness of the workshop. The first is an extension of the expansion exercise; the leader asks participants who feel they have a good outline of the taxonomy's use in a "different" area to share it with others by placing it on a blackboard, overlay, or chart and describing it to the group. Only after all volunteering participants have made their presentations may group members ask questions and offer suggestions. If this were allowed during or after each presentation, some who would volunteer would be frightened of and, thus, would not reach the dissemination stage. At the end of this session, workshop participants should be informed that those who "performed" had achieved the dissemination level.

The second activity is one that promotes more creativity among participants, demonstrates achievement of the dissemination level and, furthermore, offers participants ideas for presentations to other groups. Through any acceptable method, the participants are divided into groups of three. Each group receives copies of all materials used previously and the steps followed and is instructed to design a 20–30-minute workshop for "teaching the taxonomy." When they finish this, each group is provided time to practice, to make their presentation, and to be critiqued by the others.

## 6.0 REVIEW SUMMARY EVALUATION

At this time, leaders should summarize all that has been presented in this overview session. Leaders should recall what has been done and the points that needed clarification and should provide the participants an opportunity to evaluate candidly the entire workshop. Resource leaders should be made aware in some anonymous form of whether the objectives have been achieved and of participants' reaction to the process. Participants should now complete a suitable evaluation form, developed and provided at the start of the workshop. This evaluation should help the resource leader and the participants plan the next series of workshops at the participation level. The importance of this initial workshop cannot be overemphasized, for it provides the foundation upon which subsequent presentations are based and shortens the time necessary before participants can use the taxonomy independently to improve the teaching and learning environment.

## PARTICIPATION

The central purpose at this level is to enable participants to use the taxonomy for planning, implementing, and evaluating student learning. Workshops now should focus on the organization, the implementation and the evaluation of a curriculum unit. By the time they have achieved this taxonomic level in staff development, participants can be expected to be fully aware of the taxonomy and can be introduced to coding and to curriculum unit development. Assuming these expectations achieved, one can then begin the process of planning the participation-level workshop.

1.0 Review of the Taxonomy
2.0 Overview of Total Experience Design for Curriculum Development (Exposure)
3.0 Expanding the Total Experience Design (Participation)
4.0 Steps for Unit Development (Identification)
5.0 Participant Unit Development (Internalization)
6.0 Unit Presentation (Dissemination)

Each of these elements is a part of the linked objectives for in-service and professional development at the participation level. The leader must plan carefully with particular emphasis on Level 4.0 because subsequent-level workshops are contingent on the successful development of a curriculum unit. It is urgent that the resource leader should have gone through the curriculum development process thoroughly in order to make it easier to explain the process to workshop participants. Each element needs to be considered as follows:

*1.0 REVIEW OF THE TAXONOMY*

The first purpose of the review is to bring all participants to the same level of understanding regarding the taxonomy. When leaders find that some are not at the desired level by the end of the review, additional or brief individual instruction may be necessary. The workshop plan should include alternatives to meet this possibility. The second purpose of the review is to prepare the group for the work ahead. A brief review of the central purpose at this level refreshes the memory and establishes a set for the main action—indeed a vital and important element of any workshop. Here participants must be made aware that by the end of this component, they should have a tangible product. Each member should have completed all or a major part of a learning sequence or a teaching unit. This is the purpose of the participation-level workshop and, thus, it should be planned for taxonomically, step by step.

## 2.0 OVERVIEW OF THE TOTAL EXPERIENCE
## DESIGN FOR CURRICULUM DEVELOPMENT

This is a brief but necessary step that provides participants with the total picture of a curriculum development. It is essential if those present are being trained to develop new curricula for their school districts even if they work only at the unit level in major goal achievement. The Total Experience Design chart found at the end of Chapter 4 should be presented with the explanation that it is a basic framework from which instructional systems can be developed. The leader should emphasize that it presents the basic essentials in such a way that it ensures not only a logical progression through all levels of the taxonomy but also, upon completion, achievement of the curricular goals. The leader should explain briefly each box in the chart, using an overhead projection or blackboard diagram of the chart. Upon completion, each participant should receive a handout copy.

## 3.0 EXPANDING THE TOTAL EXPERIENCE
## DESIGN

At the program level the curriculum design includes the overall goals and several levels of objectives. Statements establishing a need or justification for the program often precede these goals. The leader will find cause to refer back as this workshop progresses and thus should briefly discuss examples of these statements, goals, and subgoals with the participants. The leader should remind the participants that the goals themselves state in general terms what the curriculum is designed to reach and that the subgoals further expand the program and clarify areas of focus. The leader may use the examples and material provided in Chapter 4 to explain each descending step of the design. Of greatest interest to the participant at this time will be the taxonomic objectives. Leaders should pay particular attention in their presentations to how these clarify, at the program level, not only the way the program objectives are to be achieved but also the relationship between the objectives, the content, and the process.

Since the central purpose of this workshop is to use the taxonomy at the classroom level, once participants group the need for the higher levels of goals and objectives, leaders should have the group explore a completed unit in order to familiarize them with the format and to set the stage for participants' questioning and understanding the need for each part. The purpose now is to expand their data base by relating the format to other aspects of the taxonomy and by bringing in other curriculum and learning theory resources. The leader is now the catalyst who sees the interaction and encourages the participants to develop first a unit together and then one of their own.

## 4.0 STEPS FOR UNIT DEVELOPMENT

The participants actually begin the process of unit preparation. The steps in Chapter 4 provide the text for workshop leaders. We have found it profitable to keep the group working as a unit for Steps 1–4 and as individuals for the remaining steps. This provides each participant with a model yet allows for a variety of individual approaches to the achievement of unit objectives. Probably one of the most interesting and reinforcing interactions comes about as participants share their differing activities and strategies for attaining the same objective.

## 5.0 PARTICIPANT UNIT DEVELOPMENT

This is the heart and circulatory system of the participation-level workshop. Here participants must decide which unit they wish to prepare. To be successful, this must be a participant's and not the leader's decision. This must be emphasized since the resultant work will be used in subsequent workshops to increase teacher effectiveness. All steps, from the development of a rationale and goals for a unit through the citing of resources, must be carried out at this time. As the sustainer, the leader must now encourage the participant to probe deeper into each step, to explore and compare available resources, and to select appropriate evaluation strategies. Furthermore, participants must be motivated to visualize the resultant unit in an actual setting, must be induced to modify where necessary, and must be prepared to implement when the opportunity arises.

## 6.0 UNIT PRESENTATION

As in the earlier workshop, leaders ask participants who feel they have a good unit to allow it to be reproduced for other members of the group. Participants are also encouraged to volunteer to summarize the steps they took in creating their unit and to advance their theories on how to teach the process to others. At this point they have completed the participation stage.

Participants now have reached a point where they have a basic understanding of the taxonomy and of its uses. They are now ready to interact with the taxonomy in a more personal context to strengthen their insight into how it can be used to improve their day-to-day contact with the learning situation.


## IDENTIFICATION

The major focus at this level involves process evaluation through coding, analyzing, and interpreting classroom interaction in the teaching–learning act. When planning and implementing this component of a taxonomically sequenced in-service program, the leader should remember that this is basically a *doing* component. Participants have understood the functional use of the taxonomy and have prepared a curriculum unit for their own classroom. They now have curricula that they prepared and that they plan to use in their own situation. This in-service component should therefore consist of several sessions spread over a period of time long enough for the people involved actually to test their unit. Each session should be planned carefully and should have certain expectations. Attention should be paid to the following:

1.0 Review of what has occurred previously and introduction of what is to take place (Exposure)
2.0 Discussion of specific processes of coding, analyzing, and interpreting material (Participation)
3.0 Coding application and group analytic and interpretive review (Identification)
4.0 Preparation of an individual brief for professional improvement, based on the analysis and interpretation of coding (Internalization)
5.0 Implementation and demonstration of change to students and peers (Dissemination)

These are the five steps leading to the fruition of the identification level of an overall in-service plan. This level, like the others, is also taxonomically sequenced. In essence, the planners can deal taxonomically with the completion of any in-service segment or program, whether it is a part of a whole or the whole itself. Each of the preceding steps has its own sequence and structure. For each part there are some specifics.

### 1.0 REVIEW OF WHAT HAS OCCURRED AND INTRODUCTION OF WHAT WILL TAKE PLACE

The leaders and the participants again need this review to establish a common basis of achievement and understanding. It can be a formal or an informal pretest. It may also be a simple straightforward review or an informal sharing. No matter what the format, it should be accomplished in as brief a time as possible. Introduction now includes group goal setting, a sequential presentation of tasks and of the timeline, and a discussion of each task as an orientation for the activities ahead. This leads logically to the next step.

### 2.0 DISCUSSION OF SPECIFIC PROCESSES OF CODING, ANALYZING, AND INTERPRETING MATERIAL

At this point each task is discussed thoroughly. Every participant should know the processes when this stage of the in-service is completed. It is important for the leader to be aware of the participant's level of understanding in order to conduct discussion and various other kinds of leader–participant, or participant–participant interaction. Small group sessions and individual conferences can help in this process. The participant should have some pretaped lessons for a practical activity that teaches coding. Likewise, materials should be on hand to acquaint participants with the steps involved in analysis and in interpretation. Upon completion of this step each participant should be able to code independently and to analyze and interpret a classroom learning session. The leaders should be prepared to assist and to support participants in this effort. This segment of the identification in-service component can be accomplished in one setting.

### 3.0 CODING APPLICATION AND GROUP ANALYTIC AND INTERPRETIVE REVIEW

In this segment, which may be spread over several sessions, the participant uses the knowledge of the process learned at the participation level and applies it to classroom teaching. The participant codes, analyzes, and interprets every tape made. Group sessions can be based on the individual experience with coding, analysis, and interpretation. In these sessions, the participant can gain confidence in the process and can begin to understand ways and strategies for making positive changes in the teaching–learning acts in the classroom. Upon completion of these sessions, the participant will feel enough self-confidence in the ability to code, analyze, and interpret to feel the need to begin an individual brief for professional change, and steps can be taken to prepare that brief.

## 4.0 PREPARATION OF AN INDIVIDUAL BRIEF FOR PROFESSIONAL IMPROVEMENT BASED ON THE ANALYSIS AND INTERPRETATION OF CODING

A brief of this nature can take almost any form ranging from the most informal kind of "I need to . . ." to a formal systems approach with column heads like "What is," "What ought to be," "Proposed activities," and "Evaluation of changes." The important element is that participants recognize the need for improvement and write a statement on what that improvement will be and how they will achieve it. Any recommended change should lead to greater student learning. Rather than the leader pointing out the improvement needed, it must come out in student performance. We contend that educators should be on guard against doing something simple because it is new. The purpose is to improve what is, providing consistent and lasting growth. When this step is completed, the teacher should be ready to implement the improvement through one in-service session.

## 5.0 IMPLEMENTATION AND DEMONSTRATION OF CHANGE TO STUDENTS AND PEERS

This is the evidence of improvement in teaching behavior and in working with students. As there is an obvious overt change, this is at the dissemination stage in a taxonomic sequence. This implementation requires three elements that the in-service leaders should plan. These elements are

- The improvement noted by the students and the teacher in classroom interaction

- The improvement that the teacher's peers share
- The results of improvement reflected in augmented student learning

The first of these is probably the most rewarding as the learners and the teachers grow together in the teaching–learning act. It is here that the real reward for the teacher comes as recognizable growth occurs and as students learn with more efficiency and with greater interest and enthusiasm. The second of these elements is the manner in which the teacher participant shares, tells, or teaches peers the professional dimension of the experience. This communicating, along with demonstrated learning improvement, is the culmination of this stage of in-service. As these are shared, the participant begins to move from element dissemination to overall internalization of the in-service process.

*INTERNALIZATION*

The major pursuit of the internalization stage is to manifest improved professional behavior in planning, implementing, and evaluating the teaching–learning act. The participant has achieved, has planned, and has made changes. Now the leaders play the role of sustainers. Their charge is to encourage, to praise, and to support the manifested change in the teacher's actions. When a teacher recognizes development and the students show real growth, the teachers voice relaxes (in class interaction) and learners are thoroughly motivated to move quickly to the higher levels of taxonomically planned curricula. The participant now plans consistently around taxonomically sequenced activities, implements them and evaluates them. The participant knows improving patterns of growth and the process of growth is apparent at this step, as self-evaluation through coding, analysis, and interpretation become a constant rather than merely a learned skill. This is the achievement level toward which the in-service has been planned. The planned objectives are met and manifested. The in-service program has achieved its ultimate purpose when the teacher reaches this level of internalization. There is, however, one additional diminsion to the in-service, the dissemination of internalized learning to others. Therefore, any ultimately successful in-service program will include plans for a dissemination opportunity for its participants.

*DISSEMINATION*

To keep something good to oneself is narrow and unprofessional. Once past the level of internalization, the teacher becomes involved in the process of helping others attain similar professional growth and development. In terms of the in-service program, a new cadre of leaders has been developed and an ever-widening circle of professionalism achieved. Though this process may seem relatively easy after the internalization in-service workshop, in-service leaders must still critique the preparation and the implementation of the dissemination stage. Although the original leaders are not in charge per se, they are assisting, critiquing, and evaluating the performance of the new presenters. The leader's tasks are to hone the new in-service leaders, to act, in a sense, as master trainers. The in-service program has come full circle.

## A PATTERN FOR CONDUCTING OTHER
## IN-SERVICE PROGRAMS

In the previous section, the focus was the improvement of teacher performance in all teaching–learning situations. The pattern for developing effective presentations in other areas, whether in-service, briefings or problem-solving sessions, still holds. Following the five taxonomic steps in these instances will be an effective in-service, since all steps will be included. Second, there will be an in-service guide that can serve as a basis for continued planning, and lastly, the leader does not have to rely on anonymously written instruments, which at best are suspect, to measure in-service success, but can determine this from the effectiveness demonstrated at the dissemination stage.

The process to be followed is much simpler than that noted in the previous sections. It must, however, be followed closely to ensure that the purpose of the meeting, the workshop, or the briefing is achieved. In our experience, the purpose should be one that results in voluntary action on the part of the participant, and, as already mentioned, this is the dissemination level. The use of the taxonomy in this instance is best illustrated by presenting our steps followed by a brief discussion of each. The presentations selected are concerned with concept and with curriculum and staff development as well as with the community briefing. Although only major categories are used, it is frequently useful to reexamine the subcategories of the taxonomy for additional clues for workshop strategies at each level.

## CONCEPT DEVELOPMENT

The popular term "individualization" has a variety of interpretations ranging from one-to-one instruction to personalized learning. Few publics, either inside or outside the field of education, have taken the time to determine what it really means to them in terms of their local educational system. The purpose of this concept development workshop, therefore, was to have educational or community groups clarify for themselves what they mean when they use the term. The steps followed were as follows.

### EXPOSURE

The workshop leader gave a 40-minute speech entitled "Erasing the Mystique." This presentation used graphics and overlays and reviewed educational trends, types of individualization, and the question, "Why individualize?"

### PARTICIPATION

Each participant received a paper with 12 statements regarding individualization, with which he or she was to agree or disagree. This is presently a popular technique that can be used in a variety of ways to help participants explore the several facets of any conceptual term.

### IDENTIFICATION

During this stage the participants formed small groups where they shared responses to the 12 statements. These groups are most effective when group participants come from different orientations. When individuals disagree, they further explore the statement together and, if they do not reach a common agreement, they rewrite the statement to reflect the consensus of the group.

## INTERNALIZATION

The workshop leader then brought the groups together to compare responses to each statement. When groups differ, they evaluate each other's responses and try to understand the others' reasoning. The leader did not seek total group consensus before moving on. The purpose is to cause everyone to reexamine their position up to this point.

## DISSEMINATION

This is the action level. When the workshop was conducted for a particular group, such as a parent council, work on a position paper was suggested. When representatives from several different groups were involved, participants produced their own written expression or outline and presented it to the total group either as a brief speech, as part of a panel, or as a written article.

## *CURRICULUM DEVELOPMENT*

The purpose of this curriculum workshop was to develop in an organized manner activities to reinforce higher level thinking skills. Prior to the meeting, the planners selected standards of proficiency, performance indicators, and criteria for activities and teacher developers. The 3-day workshop was organized in the following fashion.

## EXPOSURE

The leader reviewed standards and indicators, discussed the criteria that resultant activities were to meet, and presented and evaluated a sample activity against the criteria.

## PARTICIPATION

The group selected one thinking skill and participants walked through the activity development process together.

## IDENTIFICATION

Each participant developed one activity for an assigned standard and indicator. Throughout this part of the workshop, the leader was continuously available for reactions and for suggestions on needed additions, deletions, and changes. Participants selected a time at which to share both their progress to date and the rationale for the activity. This provided the leader a necessary opportunity for supportive feedback in a total group situation.

## INTERNALIZATION

Participants selected items (indicators) for activity development from the list provided and developed the necessary activities on their own. They were free to meet with the leader at any time for this purpose.

## DISSEMINATION

Upon completion, participants critiqued the activities developed and filled out two evaluation instruments, one on the workshop itself and one rating each activity's classroom effectiveness.

## STAFF DEVELOPMENT

The goal of the workshop described in the following was to enable the participant to conduct a similar in-service program introducing teachers to a new enrichment curriculum guide based on the experiential taxonomy. Thus, its purpose was to provide school representatives with a thorough insight into the development of the guide and a technique to use in their own presentations to their respective building staffs. It was leadership training designed to multiply the effectiveness of the initial workshop. The following procedure was used.

## EXPOSURE

The leader briefly explained how the new curriculum fit into the existing course of study, what it was, and how it was different. Since the curriculum guide was based on the Total Experience Design, an overhead projector presentation was used to illustrate the experiential taxonomy, its five major categories, the definitions of "experience," the Total Experience Design, a brief expansion of the major categories, and the "musical walk-through" noted in an earlier chapter.

## PARTICIPATION

Under the leader's guidance, participants walked through the guide and were shown its parts and the units at the beginning, the intermediate, and the advanced levels. The parts of a unit were discussed (i.e., goals, rationale, objectives, learning principles, and the various strategies). Participants then identified many examples of how and when to use the guide to supplement the regular program.

## IDENTIFICATION

By small groups, participants searched through the guide to find specific units at their teaching level

- for teaching thinking skills
- that appeared to avoid superficiality
- that illustrated diversity in our culture
- for multigrade situations
- for slow learners
- for gifted students

## INTERNALIZATION

The leader listed the selected units by these descriptors and by teaching levels on the chalkboard. When groups differed, they reanalyzed their selections and decided which was better and why.

## DISSEMINATION

When groups differed as to the best unit in a descriptor category, each sought to convince the "other" group at committee tables. Following this activity, participants developed and demonstrated the procedure they would follow in introducing the guide to their school staffs.

## COMMUNITY BRIEFING

The briefing is a simplified version of the staff development workshop, designed both to inform participants and, hopefully, to multiply the effectiveness of the meeting by reaching others who have not attended. The specific briefing that follows was designed to inform community leaders about a new program in area schools. The palnners sent invitations and acceptance notices to identified leaders and made follow-up telephone reminders.

### EXPOSURE

The coordinator welcomed participants, identified presenters, and gave an overview of the program. Although brevity is important, the quick overview is necessary since it initiates the participant in what is to follow and simply defines the focus.

### PARTICIPATION

The coordinator presented both visual and oral examples of the program's growth and of its major components, with careful attention given to defining educational expressions in lay terms. Sample packets of materials including a lesson, a placement test, a criterion test, and a student skill profile card were passed out, identified, and explained. This prepared participants further for the more in-depth presentation to follow.

### IDENTIFICATION

This was a walk-through presentation by the program's developers, using overlays, slides, and charts that covered the program materials while comfortably acquainting participants with goals, objectives, and teaching and evaluation strategies. The presenters initiated interaction with the participants, using choice questions until the participants assumed the questioning role.

### INTERNALIZATION

To expand their understanding, the participants visited the classrooms involved in the program. The participants were encouraged to observe, to talk to students and teachers, and to move from room to room.

## DISSEMINATION

Participants were encouraged to complete a "briefing evaluation" form that included space for names of others who should see the program in action. Since the intent was to reach others through participants, each received copies of a single sheet of "brief notes" that they could discuss with others, hand out, pin on bulletin boards, or have accessible at their place of business.

## SUMMARY

In this chapter, two types of in-service programs using the experiential taxonomy have been presented. The first is extensive and is highly recommended for professional curriculum consultants, for in-service leaders, and for master teachers because it immerses them in a procedure for improvement and expands their conceptualization of the educational experience itself. Whether the taxonomy is used overtly or not, participants will find they have added a heretofore undiscovered dimension to their perception of the teaching–learning act. The second use is easily followed and may be used by both the in-depth students and those that choose only to hover around the periphery yet want to improve their programs systematically.

# 12 | The Taxonomy in Retrospect

## OTHER USES OF THE TAXONOMY

In this book we have discussed at some length the utility of the experiential taxonomy in education. We have discussed use as a basis for curriculum development, for professional growth, for self-evaluation, and for in-service programming, with some specificity and through examples, data, and classroom functions. Yet there are innumerable other functions for the experiential taxonomy that should be investigated, used, and evaluated. In education one thinks, among other things, of developing the taxonomy as a more specific in-service guide for teacher development and for augmentation of the process of coding, analyzing, and interpreting the teaching–learning act through the Experiential Taxonomy Teaching Strategies Coding System (ETTSCS). Furthermore, applications of the taxonomy to specific content fields and to the development of courses of study for colleges, universities, and school districts can be functions for the experiential taxonomy.

Since the experiential taxonomy is a five-category encapsulation of human activity, it can be effective in almost any enterprise. One thinks immediately of educational enterprises in private and parochial schools, in business and industry, in church school, and in other areas. Whenever there is a need for instruction and learning, the experiential taxonomy can be a viable instrument for organizing, implementing, and evaluating such enterprises. The experiential taxonomy can be the organizing model for any type of instructional process. Even in small groups or in one-to-one conferencing, the experiential taxonomy can be a pertinent format for development of sequences and strategies that will meet group and individual counseling needs. The counselor can profit greatly from the use of the experiential taxonomy.

One of the strongest points about the experiential taxonomy is its flexibility. If a particular sequence or learning activity does not succeed or come to fruition, the teacher or user of the taxonomy who has some familiarity with it can readjust and replan the experience toward the adjusted objective. Any experienced teacher or instructional leader knows that as a lesson or learning activity progresses, there may be input from learners that could change the direction of the lesson or of the activity. With preplanned experiential taxonomy sequenced learning materials the teacher is in a very good position to make quick adjustments more easily than if the learning sequence were not carefully planned. Additionally, with a format like the experiential taxonomy, one can scrap a plan and start anew without the great expenditure of time involved in a system less sequential or without an intrinsic relationship to teaching strategies, to learning conditions, and to planned taxonomic activities, learner behaviors, and evaluation. With taxonomic planning one can be flexible and can still retain basic goals while adjusting specific objectives.

The taxonomy also has a place in the planning of any human interaction. One such example for use of the taxonomy is in political campaigns. With careful taxonomic planning one could prepare a sequential series of activities that would move listeners and voters along the taxonomic sequence to a particular voting position. In such a program, the essential element begins with a strong group of leaders who develop activities to move voters from the exposure level into an identification with (or against) a candidate. It is at the identification level that the voter usually casts his ballot. Internalization and dissemination are stages that a worker achieves, but the voters themselves need not necessarily achieve them. Careful planning with specific strategies marked out and defined could help push the voter to identification and to the casting of a ballot in a particular manner.

An equally viable area for using the taxonomy is in the advertising field. Here the sequence follows similarly to the political sequence except that there is a follow-up at the internalization level for people to keep buying the product or doing certain things. Likewise, at dissemination, there is a need for testimonial participant interaction. The use of the taxonomy can enhance the planning and implementation of advertising programs.

A dimension not discussed in either the political or the advertising uses of the experiential taxonomy is the use of media. Citizens in industrialized countries have long been consumers of the media message but, generally, have never assessed commercials, newscasts, or entertainment programs. There is a need to help consumers evaluate the message and the sequence of these in terms of the taxonomy. This could be also an exercise in preparing advertising.

Perhaps, too, one can assess the live drama of the stage. What were the unconscious techniques, for example, of Sophocles as he moved his audience to genuine catharsis of an internalized attitude in his *Oedipus the King*? How, too, does Shakespeare do this through his character Hamlet? The taxonomy can prove to be an effective instrument of literary criticism and media analysis.

One can even, with the organizational impetus of the experiential taxonomy, assess and analyze the structure, the sequence, and the strategies involved in a written creative effort. In effect, any subjective communication written for a specific purpose can have its objectives identified and its organization analyzed and interpreted in terms of the experiential taxonomy. In this way a new dimension in literary criticism and insight into the creative act may be available for professionals in the field.

Still another area, which has been referred to only in passing and in which the experiential taxonomy may be most appropriate, is individual counseling. This sensitive and highly charged interaction can be organized and sequenced using the Experiential Taxonomy to plan the counseling act. For example, when an individual seeks or is sent to a counseling situation, either the counselee or another individual sees areas where help is needed. The counselor's task is to help the counselee identify the area of need, plan new behavior patterns, try them out, internalize them, and live them. A competent and careful counselor will have the internalized goals well planned in mind as the sessions begin and, subsequently, will guide the counselee to the internalization level. The counselor needs to remember also that there is an identifiable taxonomic context to failure. One cannot fail at exposure because this is a time when the experience is being presented. An experience can be rejected at this point as it can be at participation. Here one says, in effect, "I won't." From this there may come a sense of frustration that becomes progressively stronger as the taxonomic sequence progresses. Failure occurs at the identification level when one says, "I can't." This sense of failure can, of course, result in other complications if it is constant. The counselor should keep in mind that this taxonomic level of an experience requires great planning and careful guidance to move the counselee to the success of the internalization level.

There are, as well, numerous other applications for the experiential taxonomy. It is a useful instrument for planning, analyzing, and evaluating any human experience. It is psychologically sound and educationally viable and is an appropriate tool for evaluation and for research. With these qualities, it can be used in any context of human experience.

Human experience is a wonderfully vibrant thing. Although we posit that much more can be learned about it through this taxonomy, it remains a constant that human beings are indeed unique and that each experience has a unique development. No matter how scholars study individual experience, research it, code it, analyze it, and interpret it, the experience remains uniquely and essentially human. That must be and must continue as the constant in our search for human integrity and for human dignity.

## REVIEW AND SUMMARY

In this book we have taken a substantial look at human experience. We have seen human experience as a process. This process has been defined taxonomically, marking the experiential taxonomy as unique. With experience defined taxonomically, it followed that learning, too, has an intrinsic taxonomic structure. Showing learning in this taxonomic process, the problem of developing sequenced objectives and a curriculum development system could be addressed. This step-by-step process for building curriculum is called the Total Experience Design. In any experiential learning, this curriculum structure is keyed to learning because it combines for the teacher a sound theoretical base with specific practical applications. As the instructor masters the Total Experience Design, the ability to help students learn is enhanced. Central to the development of this curriculum, a number of theoretical components were taxonomically defined, explained, and applied to the Total Experience Design. Among these were creativity, problem solving, critical thinking, learning principles, and teaching strategies. We found, as the model developed, that instead of one teaching role model, there are five roles, one for each taxonomic category.

Curriculum development and implementation is not, however, the only component of teaching and learning. There are two other notable aspects, evaluation and professional development. The Experiential Taxonomy Teaching Strategies Coding System was developed for evaluation. Essential to this system are the coding, analysis, and interpretation of the teaching–learning act. With the Total Experience Design we outlined an in-service program for promoting teacher change. Material has been included for the development of a self-directing in-service program to augment student learning and to bring about positive teacher change. Finally, uses of this taxonomy in fields other than education have been suggested. We have demonstrated that experience is a process and that the taxonomy can help in learning about that process, whether in the field of education or any other field of human activity.

With only the beginning of a continuing process of research and development, readers are encouraged to use this work to plan, implement, and evaluate experiences in the classroom or in some other context. We have presented a frame of reference for this investigation, as well as a point of view, an approach, a format, and the know-how to plan, implement, and evaluate. We feel that perhaps this exploration, research, and evaluation will throw new light on human experience. The more one works with this taxonomy, the more insight one will have into its multiple uses.

# Bibliography

Alschuler, A. S. *Developing achievement motivation in adolescents*. Englewood Cliffs, New Jersey: Educational Technology Publications, 1973.

Ausubel, D. P. *Educational psychology: A cognitive view*. New York: Holt, Rinehart, and Winston, 1968.

Basic Skills Research Grant Announcement, Spring, 1976. Washington, D.C.: The National Institute of Education.

Berelson, B. & Steiner, G. A. *Human behavior: An inventory of scientific findings*. New York: Harcourt, Brace, & World, 1964.

Bloom, B. S., Englehart, May D., Furst, Edward J., Hill, Walker H., & Krathwohl, D. and R. (Eds.), *A handbook of educational objectives: The cognitive domain*. New York: David McKay, Inc., 1964.

Bloom, B. S., Hastings, T. J., & Madaus, G. F., *Handbook on formative and summative evaluation of student learning*. New York: McGraw-Hill Book Company, 1971.

Bruner, J. S. *Toward a theory of instruction*. Cambridge, Massachusetts: The Belknap Press, 1964.

Carin, A. A., & Sund, R. B. *Developing questioning techniques: A self concept approach*. Columbus, Ohio: Charles E. Merrill Publishing Company, 1971.

Dichter, E. *Motivating human behavior.* New York: McGraw-Hill Book Company, 1971.

Eagan, K. How to ask questions that promote higher level thinking. *Peabody Journal of Education,* April 1975, pp. 228–234.

Emmer, E. T., & Millett, G. B. *Improving teaching through experimentation: A laboratory approach.* Englewood Cliffs, New Jersey: Prentice-Hall, Inc., 1970. P. 159.

Gage, N. L. Paradigms for research on teaching. In N. L. Gage (Ed.), *Handbook of research on teaching.* Chicago:Rand McNally, 1963. Pp. 94–141.

Gagné, R. M., & Rohwer, W. D. Jr. "Instructional psychology." *Annual review of psychology, 20, 1969,* 381–418.

Glaser, R. Psychological bases for instructional design. *AV Communication review, 14,* 1966, 433–449.

Gordon, I. J. (Ed.). *Criteria for theories of instruction.* Washington, D.C.:Association for Supervision and Curriculum Development, 1968.

Gordon, W. J. J. *Synthetics.* New York:Harper & Row, 1961.

Haber, R. N. *Current research in motivation.* New York:Holt, Rinehart, and Winston, Inc., 1966.

Harrow, A. J. *Taxonomy of psychomotor domain: A guide for developing behavioral objectives.* New York:David McKay, Inc., 1972.

Helder and Piaget. *The growth of logical thinking from childhood to adolescence.* New York:Basic Books, 1958.

Krathwohol, D., Bloom, Benjamin S., & Masia, Bertran B. (Eds.), *A handbook of educational objectives: The affective domain.* New York:David McKay, Inc., 1968.

Phillips, J. L. Jr. *The origin of intellect: Piaget's theory.* San Francisco:W.H. Freeman and Company, 1969.

Raths, L. E., Wasserman, S., Jonas, A., & Rothstein, A. M. *Teaching for thinking: Theory and application.* Columbus, Ohio:Charles E. Merrill Publishing Company, 1967.

Reigle, R. P. The logical status of question. *Educational Theory,* Fall, 1975, pp. 379–388.

Rosenshine, B. Classroom Instruction. *The psychology of teaching methods. The Seventy-Fifth Yearbook of the National Society for the Study of Education.* Chicago:University of Chicago Press, 1976.

Rugg, H. *Imagination.* New York:Harper & Row, 1963.

Sanders, N. M. *Classroom questions, what kinds?* New York:Harper & Row, 1966.

Schoen, R. L. Arousing the whole mental faculty. *Today's Speech,* Summer 1975, pp. 39–43.

Servey, R. E. *Teacher talk: The knack of asking questions.* Belmont, California:Fearson Publishers, 1974.

Shaftel & Shaftel. *Role playing for social responsibility.* Englewood Cliffs, N.J.:Prentice-Hall, Inc., 1967.

Sharp, B. B., & Weldon, W. *Learning: The rythm of risk.* Rosemont, Illinois:Combined Motivation Education Systems, Inc., 1971. P. 122.

Siegel, L. (Ed.). *Instruction: Some contemporary viewpoints.* San Francisco:Chandler, 1967.

Simpson, E. *The classification of educational objectives, psychomotor domain.* Champaign–Urbana, Illinois:University of Illinois Press, 1966.

Skinner, B. F. *The technology of teaching.* New York:Appleton-Century-Crofts, 1968.

Sloane, H. N. Jr., & Jackson, D. A. *A guide to motivating learners.* Englewood Cliffs, New Jersey:Educational Technology Publications, 1974.

Steinaker, N. & Bell, M. R. A proposed taxonomy of educational objectives: The experiential domain. *Educational Technology,* January, 1975, pp. 14–16.

Steinaker, N., & Bell, M. R. An evaluation design based on the experiential taxonomy. *Educational Technology,* February 1976, pp. 26–29. (a)

Steinaker, N., & Bell, M. R. How teachers can use the experiential taxonomy. *Educational Technology,* November 1976, pp. 49–52. (b)

Steinaker, N., & Harrison, M. *Measuring experiences through the experiential taxonomy: A research study in special education.* Ontario, California:Ontario–Montclair School District, June 1976.

Steinaker, N., & Harrison, M. *Taxonomic organization for classroom heuristicism.* Ontario, California:Ontario–Montclair School District, 1977.

Travers, R. M. *Essentials of learning.* New York:The MacMillian Company, 1963.

Wease, H. Questioning: The genesis of teaching and learning. *High School Journal,* March 1976, pp. 258–265.

Webster's New World Dictionary of the American Language, Second College Edition. Cleveland, Ohio:William Collins and World Publishing Company (copyright 1976).

Whitehead, Alfred North. *Process and reality: An essay in Cosmology.* New York: The Macmillan Company, 1929.

# Subject Index

# EDUCATIONAL PSYCHOLOGY

*continued from page ii*

António Simões (ed.). The Bilingual Child: Research and Analysis of Existing Educational Themes

Gilbert R. Austin. Early Childhood Education: An International Perspective

Vernon L. Allen (ed.). Children as Teachers: Theory and Research on Tutoring

Joel R. Levin and Vernon L. Allen (eds.). Cognitive Learning in Children: Theories and Strategies

Donald E. P. Smith and others. A Technology of Reading and Writing (in four volumes).

> *Vol. 1. Learning to Read and Write: A Task Analysis (by Donald E. P. Smith)*
>
> *Vol. 2. Criterion-Referenced Tests for Reading and Writing (by Judith M. Smith, Donald E. P. Smith, and James R. Brink)*
>
> *Vol. 3. The Adaptive Classroom (by Donald E. P. Smith)*
>
> *Vol. 4. Designing Instructional Tasks (by Judith M. Smith)*

Phillip S. Strain, Thomas P. Cooke, and Tony Apolloni. Teaching Exceptional Children: Assessing and Modifying Social Behavior